# *INDUSTRIAL MATERIALS*

## *VOLUME 2*

## *Polymers, Ceramics, and Composites*

*David A. Colling*
University of Massachusetts Lowell

*Thomas Vasilos*
University of Massachusetts Lowell

Prentice Hall
Upper Saddle River, NJ 07458     Columbus, Ohio

**Library of Congress Cataloging-in-Publication Data**

Colling, David A.
Industrial materials.

Includes index.
Contents: v. 1. Metals and alloys—
    v. 2. Polymers, ceramics, and composites.
    1. Materials.   2. Materials—Case studies.
    3. Manufacturing processes—Case studies.
I. Vasilos, Thomas.   II. Title.
TA403.C585   1995     620.1'1          94-10077
ISBN 0-02-323553-5 (v. 2)

Cover photo: David A. Colling
Editor: Stephen Helba
Production Editor: Mary Ann Hopper
Text Designer: Julia Zonneveld Van Hook
Cover Designer: Julia Zonneveld Van Hook
Production Buyer: Patricia A. Tonneman
Electronic Text Management: Marilyn Wilson Phelps, Matthew Williams, Jane Lopez,
    Karen L. Bretz
Illustrations: Diphrent Strokes, Inc.

This book was set in Times by Prentice Hall and was printed and bound by R. R. Donnelley &
Sons Company. The cover was printed by Phoenix Color Corp.

© 1995 by Prentice-Hall. Inc.
A Pearson Education Company
Upper Saddle River, NJ 07458

Printed in the United States of America

10 9 8 7 6 5 4 3 2 1

ISBN: 0-02-323553-5

Prentice-Hall International (UK) Limited,London
Prentice-Hall of Australia Pty. Limited, Sydney
Prentice-Hall Canada Inc., Toronto
Prentice-Hall Hispanoamericana, S.A., Mexico
Prentice-Hall of India Private Limited, New Delhi
Prentice-Hall of Japan, Inc., Tokyo
Pearson Education Asia Pte. Ltd., Singapore
Editora Prentice-Hall do Brasil, Ltda., Rio de Janeiro

# *Preface*

*T*echnology involves all phases of product development from design through delivery. Industrial practitioners must be well versed in all aspects of manufacture, including computer applications, manufacturing processes, quality control, production management, and organizational behavior, among others. Selection and processing of industrial materials are important, but students preparing for careers in industry do not need to first become proficient in materials science, as many of the good textbooks on materials available at present assume. To be successful, we feel that the authors should be specialists in materials first, but should also be well versed in product-oriented manufacturing processes, where making it right is the only practice that matters! For these reasons, this textbook is filled with case studies that illustrate industrial problems.

We also believe that teaching is important; the material in this textbook has been developed over many years of teaching industrial technology students, many of whom are now practicing in successful careers and some of whom have provided case studies that appear in the text. These students did not have rigorous mathematics backgrounds, so it was important to develop their understanding of concepts rather than their computational skills or theoretical knowledge. The thrust of this textbook is to define properties needed for applications, then relate these properties to the material properties for appropriate selection and control through processing.

It is impossible to cover all materials in a single one-semester course, yet curriculum demands do not always permit time for a two-course sequence. We have separated our treatment of materials into two volumes rather than including everything in a single cumbersome volume that might not be fully utilized. Where only a single course is required, your emphasis can be tailored to either metals and alloys

(Volume 1) or to polymers, ceramics, and composites (Volume 2), eliminating one volume or leaving it for an elective.

There is little new information provided in these two volumes—we have borrowed freely from other sources whose permission is acknowledged with appreciation. Our contribution is in the organization of the topics and their presentation in a logical fashion to establish a basis for optimum applications of materials in manufacturing.

Volume 2 is confined to polymers, ceramics, and composites. We begin the first chapter by discussing properties related to these materials and some of their applications, then we begin to build a basis for understanding the selection and control needed to satisfy the demands of these materials, beginning with brief treatments of atomic structure and bonding in organic polymers and inorganic crystalline ceramics. Chapters 2 and 3 are devoted to the structure, composition, and processing of polymers. Chapters 4 and 6 are devoted to crystalline ceramic materials and their processing. Chapter 5 covers the vitreous ceramics and their processing. In Chapter 7, the concept of composites that combine different materials and their use in engineering applications is thoroughly examined in a down-to-earth fashion. New as well as established techniques for making composites are covered in Chapter 8. Chapters 9 and 10 also deal with composites, but these are very special composites. They have been around for centuries, perhaps for millennia, yet we have only recently learned about their structures, which determine their properties. Concrete, which is the subject of Chapter 9, and wood, which is the subject of Chapter 10, lead all other materials in annual consumption.

We would like to thank our students for their inspiration and our colleagues at the University of Massachusetts Lowell for their encouragement and support. Particular thanks are due Professors R. Malloy, C. Connelly, J. Walkinshaw, and D. Leitch for reviewing sections of the manuscript. We would also like to thank those who reviewed the final manuscript: C. J. Law, Western New Mexico University; David H. Devier, Ohio Northern University; and Bill G. Cullins, Aims Community College. Finally, we could not have completed the text without the sacrifice of our families and friends, particularly our wives, Dr. Jane Dreskin and Mrs. Helen Vasilos.

# Contents

## 1

## Properties of Nonmetallic Materials and Composites   1

| 1.1 | Mechanical Properties   3 |
|---|---|
| 1.1.1 | Tensile Properties   3 |
| 1.1.2 | Hardness   8 |
| 1.1.3 | Toughness   8 |
| 1.1.4 | Stress Concentration   10 |
| 1.1.5 | Fatigue   11 |
| 1.1.6 | Wear   12 |
| 1.2 | Electrical and Thermal Properties   13 |
| 1.3 | Thermal Effects   13 |
| 1.4 | Optical Properties   14 |

Summary   15
Terms to Remember   15
Problems   16

## 2

## Polymer Materials   17

| 2.1 | Polymer Materials   18 |
|---|---|
| 2.2 | Structure of Polymers   20 |

2.3 Additives to Polymers 22
2.3.1 Improved Properties 23
2.3.2 Improved Processing or Process Control 24
2.3.3 Color Additives 25
2.3.4 Reduction of In-Service Degradation 25
2.4 General Applications of Polymers 26
2.5 Polymer Classifications 28
2.5.1 Thermoplastic Polymers 29
2.5.2 Thermosetting Polymers 35
2.5.3 Elastomers 38
2.5.4 Miscellaneous Polymers Better Classified by Shape or Utilization 43
Summary 44
Terms to Remember 44
Problems 45

# 3

# Polymer Processing 47

3.1 Processing of Thermoplastic Polymers 48
3.1.1 Extrusion 49
3.1.2 Injection Molding 51
3.1.3 Thermoforming 53
3.1.4 Miscellaneous Processes for Thermoplastic Polymers 55
3.2 Processing of Thermoplastic Polymers 63
3.2.1 Compression Molding 63
3.2.2 Transfer Molding 64
3.2.3 Pultrusion 64
3.2.4 Casting Processes 67
Summary 69
Terms to Remember 69
Problems 70

# 4

# Crystalline Ceramic Materials 71

4.1 Properties of Ceramics 72
4.1.1 Atomic Packing Factor 73

4.1.2    Thermal Conductivity    73
4.1.3    Mechanical Strength    74
4.1.4    Toughness    76
4.2    Thermal Shock Behavior    76
4.3    Crystal Structure    78
4.3.1    Pauling's Rules and Coordination Number    79
4.4    Technical Ceramic Compounds    81
4.4.1    Compounds with the NaCl Structure    81
4.4.2    Compounds with the Fluorite Structure    81
4.4.3    Compounds with the Perovskite Structure    83
4.5    Traditional Ceramic Compounds    84
4.5.1    Clays    84
4.5.2    Diamond and Graphic Forms of Carbon    84
4.5.3    Ferrites    87
Summary    87
Terms to Remember    88
Problems    88

# 5
# Glass    91

5.1    Glass Structure    92
5.2    Glass Formation    93
5.3    Glass Compositions    96
5.4    Viscosity of Glass    98
5.5    Glass Forming    99
Summary    102
Terms to Remember    102
Problems    102

# 6
# Ceramics Processing    103

6.1    Traditional Ceramics    103
6.2    Modern Ceramics    106
6.3    Raw Material Processing    107
6.4    Fabrication Methods    110
6.4.1    Casting Processes    110
6.4.2    Doctor Blade Process    112

6.4.3 Dry Processing 113
6.4.4 Isostatic Pressing 113
6.4.5 Plastic Forming 116
6.5 Densification by Sintering 119
Summary 121
Terms to Remember 121
Problems 122

# 7

# *Composite Materials 123*

7.1 Dispersion-Strengthened Composites 124
7.2 Particle-Reinforced Composites 125
7.3 Fiber-Reinforced Composites 126
7.3.1 Fiber Materials 127
7.3.2 Polymer-Matrix Fiber-Reinforced Composites 130
7.3.3 Metal-Matrix Fiber-Reinforced Composites 136
7.3.4 Other Metal Matrix Composites 139
7.3.5 Ceramic-Matrix Fiber-Reinforced Composites 141
Summary 143
Terms to Remember 144
Problems 144

# 8

# *Processing of Fiber-Reinforced Composite Materials 145*

8.1 Processing of Plastic Matrix Composites 145
8.1.1 Lay-up Processes 145
8.1.2 Compression Molding 148
8.1.3 Filament Winding 150
8.1.4 Resin Transfer Molding 151
8.1.5 Carbon Fiber-Reinforced Carbon-Matrix Composites 152
8.2 Processing of Metal Matrix Composites 154
8.2.1 Continuous Fiber-Reinforced Metal-Matrix Composites 154
8.2.2 Discontinuous Fiber-Reinforced Metals 158
8.3 Processing of Ceramic Fiber-Reinforced Ceramic-Matrix Composites 160
Summary 161
Terms to Remember 162
Problems 162

# 9
# *Concrete   163*

9.1      Types of Concrete   163
9.2      Concrete Ingredients   164
9.2.1    Aggregates   164
9.2.2    Cement   165
9.2.3    Water   168
9.3      Concrete Mixtures   168
9.4      The Strength of Concrete   170
9.5      Freeze-Thaw Resistance and Air-Entrained Concrete   174
9.6      Admixtures in Concrete   175
9.6.1    Mineral Admixtures   176
9.7      Curing of Concrete   177
9.8      Reinforced Concrete   179
9.8.1    Design Considerations   179
9.8.2    Prestressed Concrete   181
9.8.3    Corrosion of Reinforcing Steel   181
9.8.4    Fiber Reinforcement   182
9.9      Polymer Concrete   182
9.10     Design Control of Cracking   184
Summary   184
Terms to Remember   186
Problems   186

# 10
# *Wood and Wood Products   189*

10.1     Structure of Wood   190
10.2     Properties of Wood   193
10.2.1   Mechanical Properties of Lumber   199
10.2.2   Electrical, Thermal, and Chemical Properties of Wood   204
10.3     Modified Wood for Protection Against Decay   204
10.4     Wood Composites   205
10.4.1   Plywood   205
10.4.2   Particleboard   206
10.5     Paper   207
10.5.1   Separation of Wood Fibers   208

10.5.2   Making Paper   208
10.5.3   Structure of Some Paper Products   209
10.5.4   Packaging   212
Summary   215
Terms to Remember   216
Problems   216

**Glossary   219**

**Index   229**

# 1

# Properties of Nonmetallic Materials and Composites

The history of man can be measured by the development and applications of materials. It is with good reason that we refer to periods of human culture as the Stone Age, the Bronze Age, or the Iron Age because materials developments during these ages led to the production of weapons for hunting or warfare and to the production of cooking and storage utensils. Even today, our sophisticated technology is dependent on materials developments at work, at home, and at play. Enterprises such as the transportation, computer, electronics, communications, and aerospace industries are a result of our being able to study and learn about the materials needed to develop the dreams of entrepreneurs.

Our understanding of materials did not really begin until the late nineteenth century when the microscope and methods for testing materials were first developed. Up until then, materials development was purely empirical, thus limiting the technology of the time. For example, both the social and economic development of the United States were made possible by railroads. Prior to improved steelmaking by the Bessemer process, however, rails were too weak to sustain the constant travel of steam engines.

Today's materials can be classified as metals and alloys, as polymers or plastics, as ceramics, or as composites, which are combinations of different materials. Applications of these materials depend on their properties; therefore we need to know what properties are required for the application and be able to relate those specifications to the material. For example, a ladder must withstand a design load, the weight of a person using the ladder. However, the material property that can be measured is strength, which is affected by the load *and* design dimensions. Strength values must therefore be applied to determine the ladder dimensions to ensure its safe use.

The properties that we will be using throughout this textbook are those of nonmetallic materials (ceramics, polymers, and composites). They include *physical properties* such as density and melting point; *mechanical properties* such as strength, modulus, and elasticity; *electrical* and *thermal properties* such as conductivity; *magnetic properties*; and *optical properties*.

The units for measurement of properties are supposedly uniform, with the International System of Units (SI units) universally acceptable. Nevertheless, conventional usage of British units in the United States has persisted in many disciplines. This mix really does not present a problem, however, because we can readily convert to SI units when measuring in British units. Table 1.1 compares the units of measurement and lists conversion factors.

**Table 1.1**
Measurements and material properties

| Property | SI unit | British unit | Conversion factors |
|---|---|---|---|
| Length | meter | inch, foot | 1 in. = 2.54 cm = 25.4 mm<br>1 m = 39.37 in.<br>1 Å = $10^{-8}$ cm<br>1 mil (.001 in.) = .0394 mm |
| Mass | kilogram | pound mass (lbm) | 1 kg = 2.204 lbm<br>1 lbm = 453.7 g |
| Force | newton (N) | pound force (lbf) | 1 lbf = 4.44 N |
| Stress | pascal (Pa) | lbf/in.² or psi | 1 Pa = 1 N/m² = .145 × $10^{-3}$ lbf/in.²<br>1 lbf/in.² = 6.89 × $10^{3}$ Pa |
| Temperature | °C<br>K (absolute) | °F<br>°R (absolute) | °F = ⁹⁄₅°C + 32<br>K = °C + 273<br>°R = °F + 460 |

# 1.1 Mechanical Properties _____

## 1.1.1 Tensile Properties

**Mechanical properties** are associated with behavior of a design subjected to a mechanical force. For structural applications, mechanical properties are always specified in material selection. In these cases, it is typical to specify tensile properties, which simply refer to the applied forces that stretch a shape. (Compressive properties, where the applied force squeezes the shape, are always greater than tensile properties and are not usually specified.) Tensile tests are performed in universal machines such as that shown in Figure 1.1. Universal testing machines can test

**Figure 1.1**
Universal testing machine for mechanical property measurement (Courtesy of Instron Corporation, Canton, Massachusetts)

materials under shear or flexure as well. In most cases, we will specify engineering properties, which are determined from a stress-strain curve of test results.

Engineering **stress**, $\sigma$, is defined as

$$\sigma = \frac{P}{A_o}$$

where $A_o$ is the original cross-sectional area and $P$ is the force that is applied. This applied force can extend or *elongate* a material, causing a **strain**, $\varepsilon_E$, which is given by

$$\varepsilon = \frac{\ell - \ell_o}{\ell_o}$$

where $\ell$ is the specimen length after force is applied and $\ell_o$ is the original specimen length.

In contrast to most metals and many plastics that can extend or elongate beyond the elastic region, ceramic materials most often exhibit brittle failure; that is, they fracture at the elastic limit. This characteristic is illustrated in Figure 1.2 by the stress-strain curve for a typical ceramic sample. Initially, there is a large *linear*

**Figure 1.2**
Stress-strain diagram for
typical ceramic material

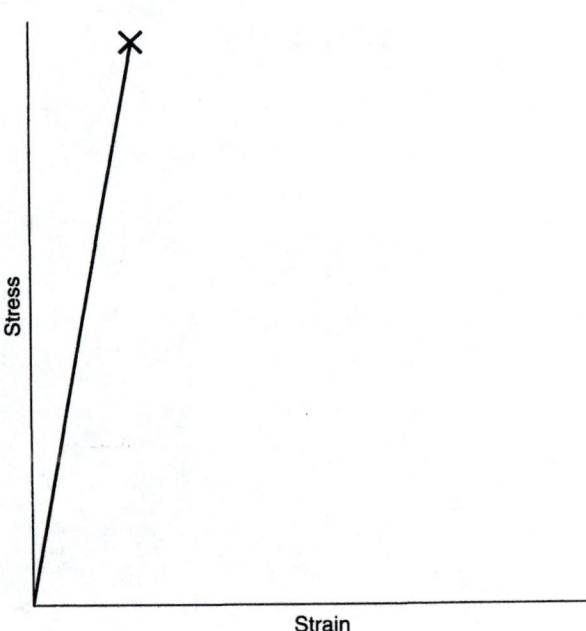

**Figure 1.3**
Stress-strain diagrams for typical plastics: (a) brittle material; (b) soft, weak material; (c) hard, tough material (Toughness is proportional to the area under the stress-strain diagram.)

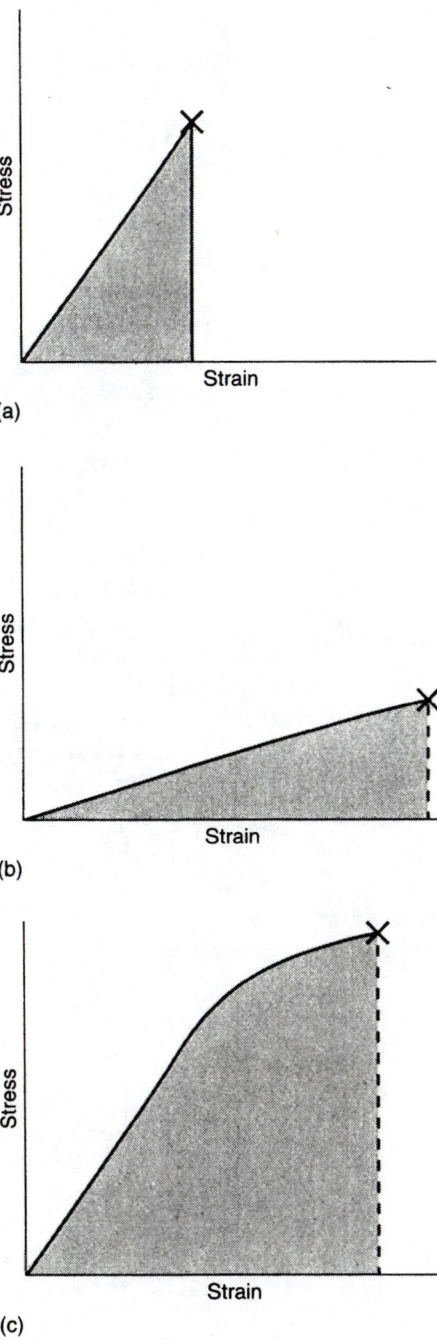

increase in stress with little strain. This is **elastic deformation** because the ceramic shape is completely recovered if the force is removed. The linear relation is described by the equation

$$\sigma = E\varepsilon$$

where $E$ is the **elastic modulus**, or Young's modulus. This equation is known as Hooke's law.

Polymer materials that are brittle exhibit stress-strain curves that are similar to those of ceramic materials, but their elastic moduli are much smaller; that is, the slope of the linear portion of the stress-strain curve is not as steep. These materials have a wide range of elastic properties, from brittle to weak and soft to strong and tough, as shown in Figure 1.3. Polymers also differ from metals and ceramics because they are sensitive to how quickly strain is applied; that is, they might be brittle when tested at high strain rates yet exhibit **ductility** at low strain rates. We call such behavior **viscoelasticity**. Noncrystalline glass ceramics also exhibit viscoelastic or semirigid behavior at temperatures above $T_g$, the glass transition temperature.

For many polymer and composite materials, flexural, or bend, testing is common. Three-point bending, which makes use of a beam sample and supports such as those shown in Figure 1.4, is conducted in a universal testing machine in **compression**. Maximum bending stress and deflection is recorded as a function of the load $P$. Flexural strength is the value of the maximum tensile stress at a specified deflection. The *modulus of rupture*, a common term describing wood strength, is the

**Figure 1.4**
Three-point flexural testing

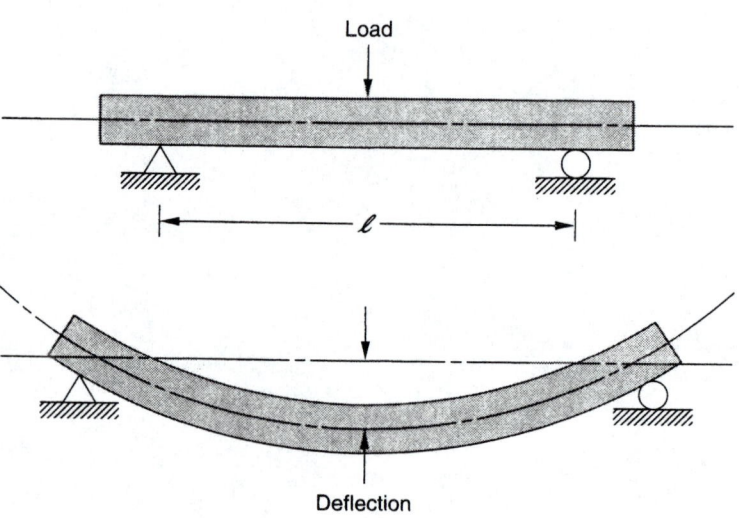

value of the flexural strength when fracture occurs. This is not a true property because the formulas are not valid once the elastic limit is exceeded, but it is an accepted criterion.

---

### Sample Problem 1.1

### Elastic Modulus for Plastic and Ceramic

The modulus of elasticity, $E$, is $55 \times 10^6$ psi for aluminum oxide and $4 \times 10^5$ psi for polyvinyl chloride (PVC).

   a.   What is the stress on an aluminum oxide rod 12 in. long and 0.125 in. in diameter that is stretched 0.02 in.? What is the force needed to stretch the rod?

   b.   What is the stress on a PVC rod 12 in. long and 0.125 in. in diameter that is stretched 0.02 in.? What is the force needed to stretch the rod?

### *Solution*

$$\varepsilon = \frac{.02}{12} = 1.67 \times 10^{-3} \text{ in./in.}$$

a.  For $Al_2O_3$

$$\sigma_E = E\varepsilon$$
$$\sigma_E = 55 \times 10^6 \text{ psi} \times 1.67 \times 10^{-3} \text{ in./in.}$$
$$= 91{,}850 \text{ psi}$$
$$\sigma_E = \frac{P}{A}$$
$$91{,}850 = \frac{P}{\pi(.125)^2/4} \quad \text{since } A = \frac{\pi d^2}{4}$$
$$P = 91{,}850 \times \frac{\pi d^2}{4}$$
$$= 1127 \text{ lb}$$

b.  For PVC

$$\sigma_E = E\varepsilon$$
$$\sigma_E = 4 \times 10^5 \text{ psi} \times 1.67 \times 10^{-3} \text{ in./in.}$$
$$= 668 \text{ psi}$$
$$\sigma_E = \frac{P}{A}$$
$$668 = \frac{P}{\pi(.125)^2/4} \quad \text{since } A = \frac{\pi d^2}{4}$$
$$P = 668 \times \frac{\pi d^2}{4}$$
$$= 32.8 \text{ lb}$$

---

### Sample Problem 1.2

### Modulus of Rupture for Wood

The modulus of rupture (MOR) is given by the outer fiber stress (a tensile stress) in three-point bending to failure:

$$MOR = \frac{3P\ell}{2wt^2}$$

where $P$ is the failure load, $\ell$ is the length between supports, $w$ is the specimen width, and $t$ is the specimen thickness.

If the MOR for eastern white pine is 8600 psi and the MOR for white oak flooring is 15,200 psi, what thickness of oak will fail at the same load as 3/4-in. pine? (Assume the boards have the same width.)

$$\text{Given } P_{pine} = P_{oak}$$
$$t_{pine}^2(MOR)_{pine} = t_{oak}^2(MOR)_{oak}$$
$$0.75^2 \times 8600 = t_{oak}^2 \times 15{,}200$$
$$t_{oak} = 0.564 \text{ in.}$$

## 1.1.2   Hardness

**Hardness** testing measures the resistance of a material to **plastic deformation**. We use it frequently as a quick method to estimate the strength of metals. The test is not used in engineering applications for brittle ceramics, however, simply because they exhibit little or no plastic deformation. Hardness testing is sometimes used for polymers, but with very different loads and indenters than those used for metals. Shore hardness, for example, uses a sharpened rod that is spring loaded, pushed against a plastic specimen by hand, and then released.

## 1.1.3   Toughness

**Toughness** is the ability of a material to absorb energy before it breaks or fractures. If we consider the polymers with stress-strain curves depicted in Figure 1.3, the toughness of the soft, weak material is higher than that of the brittle one, but the hard and tough plastic has the highest toughness. This is true because toughness is directly proportional to the total area under the stress-strain curves, an area that represents the total energy that is absorbed in fracture.

It is more common, however, to measure the toughness of a material by an **impact** test that measures the energy absorbed in fracture under an impulse load. In this type of test, we release a heavy pendulum from a known height; the pendulum strikes and breaks the sample, and then continues its upward swing. By knowing the weight of the pendulum and the heights involved, the energy absorbed in fracture can be directly measured on the machine. The most common impact tests for plastics are the Charpy and Izod tests, which use the same machine, shown in Figure 1.5, but have different specimen geometries and different impact hammers.

There is another toughness measure that we call fracture toughness, the test for which is adapted from the study of fracture mechanics. **Fracture toughness** is represented by the symbol $K_{1c}$, the stress factor to cause catastrophic failure, which is expressed in ksi in.1/2 (where ksi is 1000 pounds per square inch):

$$K_{1c} = Y\sigma_f \sqrt{\pi a}$$

where $Y$ is a geometric number near unity, $\sigma_f$ is the stress applied at failure, and $a$ is the length of a surface crack. If the flaw is located away from the surface, $1/2a$ is used in the equation. $K_{1c}$ is also affected by temperature, a phenomenon we will use to advantage in ceramic processing. It should be noted that $K_{1c}$ is appropriate only for thick sections where plane strain conditions are applicable.

**Figure 1.5**
Impact tester for Charpy and Izod testing of plastic materials (Photograph courtesy of Tinius Olsen Testing Machine Co., Inc., Willow Grove, PA.)

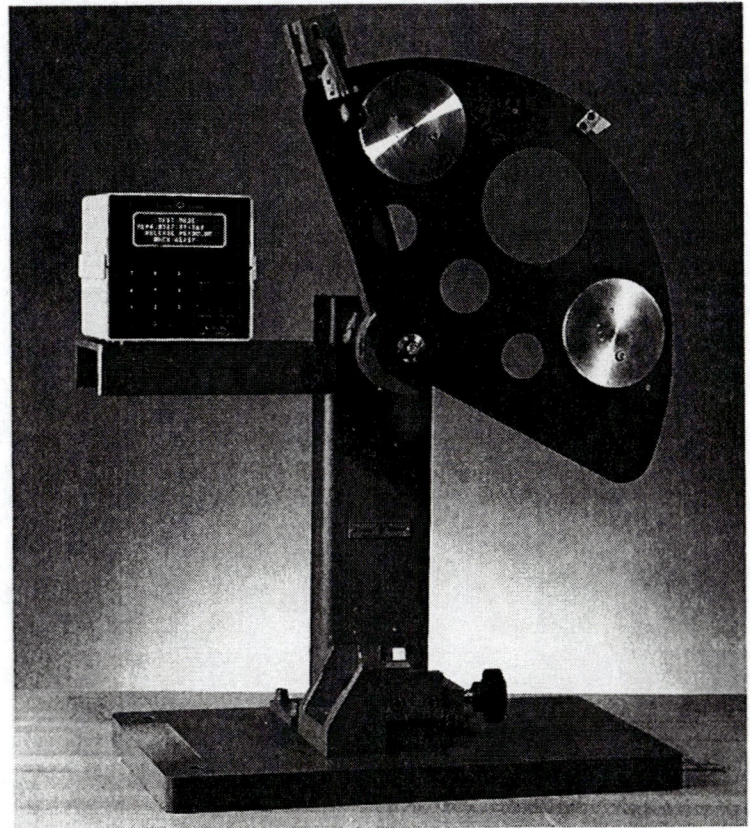

## 1.1.4   Stress Concentration

**Stress concentration** is a geometric or design factor that can lead to failure at applied stress levels much below those anticipated for failure to occur. In brittle materials, the failure can be sudden, caused by rapid propagation of a crack. However, the effects can be blunted somewhat by deformation preceding crack propagation for ductile materials. Look at Figure 1.6. The highest stress at the tip of a crack is given by the equation

$$\sigma_{max} = 2\sigma \left(\frac{c}{\rho}\right)^{1/2}$$

where $\sigma_{max}$ is the maximum stress at the crack tip, $c$ is half the length of an interior crack or is the length of an exterior crack, $\rho$ is the radius of curvature at the tip of the crack, and $\sigma$ is the applied tensile stress. We can readily see that the stress concentration can lead to crack propagation and localized failure when the bulk of the material is under a fairly low stress level.

Stress concentration occurs at drilled holes, fillets, and other design configurations, but also can result from material defects. We cannot emphasize the effects of stress concentration enough, for they cannot be ignored without serious consequences. We also must recognize that brittle materials can fail at very low applied stresses because of stress concentration; therefore, we use these brittle materials only in compression. (This fact has been recognized since ancient times when stone and brick were used by the Romans only in compression in arches in walls and aqueducts.)

---

**Sample Problem 1.3**

**Threaded Plastic Pipe**

A 2-in. diameter PVC pipe with 1/8-in. wall thickness is loaded in **tension** with a 20-lb weight. Compare the tensile stress in the pipe with that adjacent to a threaded joint where the threads are 0.053 in. deep and the radius of curvature is 0.01 in.

**Solution**

$$\sigma_{pipe} = \frac{P}{A} = \frac{20}{\pi(OD^2 - ID^2)/4}$$

$$= \frac{80}{\pi(2.00^2 - 1.75^2)}$$

$$= 27.2 \text{ psi}$$

With stress concentration,

$$\sigma_{max} = 2 \times 27.2 \left(\frac{0.053}{0.01}\right)^2$$

$$= 1528 \text{ psi}$$

---

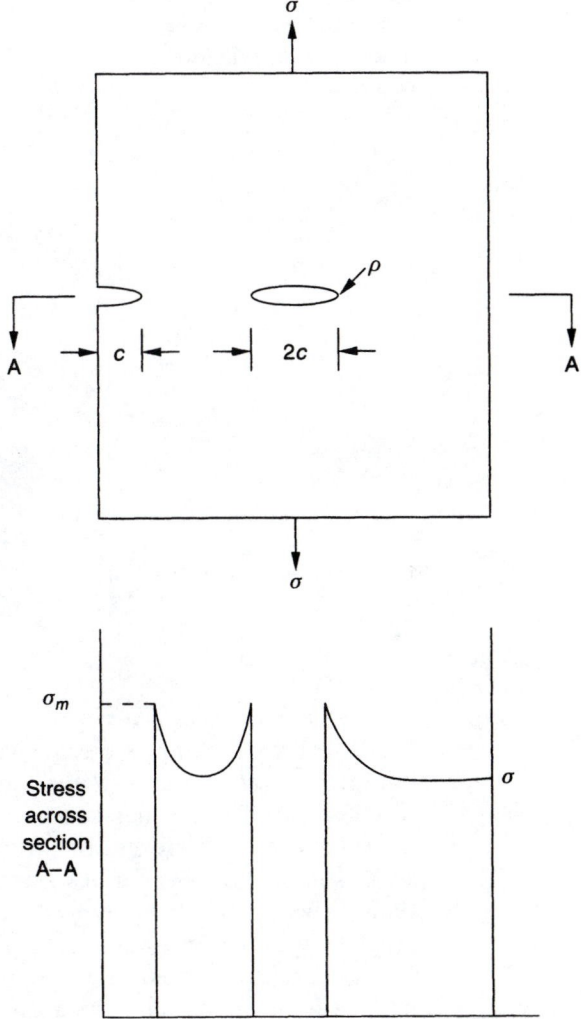

**Figure 1.6**
Stress concentration
caused by cracks

## 1.1.5 *Fatigue*

The stress-strain curve represents failure under a single applied force. What happens when we repeatedly load a material to a level below the **ultimate tensile strength**? We have applied a **fatigue** load, that is, a cyclic or intermittent load. Failure under such conditions can occur at stress levels below the ultimate tensile strength after a number of cycles.

Although fatigue failures are better known for metals, all materials can fail under fatigue conditions. For example, polymers are frequently used for containers with integral hinges, such as the one shown in Figure 1.7. Designs like this must take into consideration the fatigue properties of the specific polymer.

**Figure 1.7**
Example of plastic integral
hinge subject to fatigue
stresses

# 1.1.6   *Wear*

We have been looking at mechanical properties that can cause failure by fracture, but there are two other ways that materials can lose their usefulness—by wear or by obsolescence. We neither can nor want to control obsolescence, but we have to understand how to prevent or protect against wear. Wear is simply the removal of small amounts of a material's surface by mechanical action. Although we usually think of wear as harmful, there are useful applications such as writing with a pencil: graphite wear particles are transferred to paper by our mechanical action.

Wear actually comprises a number of different processes that take place independently or in combination. Adhesive wear that occurs when two surfaces slide across each other is the most common, but we also encounter abrasive wear, as in the writing example and in wood refinishing with abrasive sandpaper. Less common wear phenomena are corrosive wear, fatigue wear, and deformation wear. The wear process is very complex, involving many variables that cannot always be controlled. These variables include the mechanical properties of both surfaces, surface finish, contact pressure, lubrication, contaminants, and more.

Wear is of particular interest in manufacturing machinery whose surfaces move with respect to each other. When there is lubrication, wear does not occur because there is no metal-to-metal contact. However, there is always momentary contact during start-up or shutdown procedures. At those times, wear of the softer, bearing metal can occur. Polymer materials are used frequently to prevent metal wear, taking advantage of the lower friction provided by the polymer and sacrificing the less expensive bushing material to protect the machinery.

# 1.2   Electrical and Thermal Properties

Whereas we specify mechanical properties for structural applications, we must specify the **electrical properties** of materials, such as **resistance** and **resistivity**, for any applications involving conduction or insulation from electrical fields. At the same time, we will be able to estimate the thermal characteristics because thermal and electrical properties are interrelated. A good electrical conductor has to be a good thermal conductor and a poor electrical conductor has to be a poor thermal conductor with few exceptions.

In this volume, we will concentrate on materials used for insulation purposes. Both ceramics and polymers are used as insulators. One of the characteristics that is important to this application is the **dielectric strength**, often at microwave frequencies as well as at 60 hertz. The dielectric constant is a measure of the material's ability to store electrical energy without breaking down. Closely related to this is the dissipation factor, which is a measure of the power loss in the insulating material.

# 1.3   Thermal Effects

Change in temperature affects all properties to some extent. For example, electrical conductivity increases as the temperature decreases. The most common effect, however, is thermal expansion and contraction, which can lead to mechanical stress levels in materials subjected to changing temperature. In some instances, if thermal conductivity is low and thermal changes are high, failure can occur by **thermal shock**.

The stress caused by thermal expansion is given by

$$\sigma = E\alpha\Delta T$$

where $\alpha$ is the thermal expansion coefficient and $\Delta T$ is the change in temperature. Strain can be translated into stress if the object is constrained from expansion by using Hooke's law.

---

**Sample Problem 1.4**

The thermal expansion coefficient of window glass is $9 \times 10^{-6}$ in./in./°C. By adding $B_2O_3$, we can make a heat-resistant borosilicate glass, familiar to us as Corning Glass's Pyrex. The thermal expansion coefficient of Pyrex is $5 \times 10^{-6}$ in./in./°C. Assume that the modulus of elasticity of both materials is equal to $9 \times 10^6$ psi and that we have two identical rods rigidly fixed at both ends. What kind of stress and what magnitude of stress is caused by submerging the room-temperature rods (20°C) into boiling water?

### Solution

The stress will be compressive because the length of the rods would increase if free, but they are fixed.

$$\sigma = E\alpha\Delta T$$

For window glass,

$$\sigma = 9 \times 10^6 \times 9 \times 10^{-6} \times 80 = 6480 \text{ psi}$$

and for Pyrex,

$$\sigma = 9 \times 10^6 \times 5 \times 10^{-6} \times 80 = 3600 \text{ psi}$$

---

# 1.4   *Optical Properties*

Closely related to some electrical characteristics are the **optical properties** of materials. We are all familiar with the optical transparency of glass ceramics, but there are other optical characteristics that are important as well. Light, like sound, is an electromagnetic wave with specific ranges of frequencies and wavelengths. Visible light, for example, is electromagnetic radiation with wavelengths from about 0.39 to 0.77 micrometers ($\mu$m), giving us colors from violet to red. When a beam of light is directed at an optical plate (such as glass), some of the light is reflected, some is transmitted, and some is absorbed, as shown in Figure 1.8. Transmitted light is bent or refracted as well, dependent on the property we call index of refraction, $n$, which varies from about 1.5 for optical glass and polymer compositions to as high as 2.4 for diamond.

Conventional light sources give off, or emit, light waves that are randomly directed and not in phase, thus providing what we call *incoherent light*. Certain materials, called **laser** materials, emit coherent light that does not diffuse and does not lose intensity over very long distances. (The acronym *laser* means *l*ight *a*mplification by *s*timulated *e*mission of *r*adiation.) We will not study laser materials but will examine optical fibers, which are the basis for modern fiber optics communication networks. In these networks, electrical signals are converted to laser signals that travel through the optical fibers, then are reconverted to electrical signals.

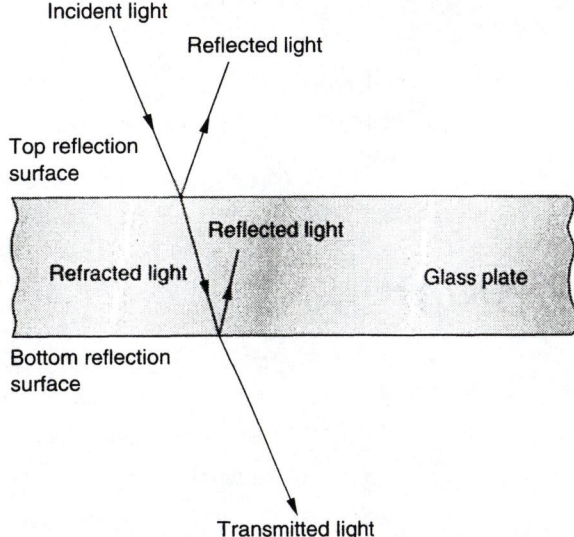

**Figure 1.8**
Reflection, refraction, and transmission of light through glass

Incident light

Reflected light

Top reflection surface

Reflected light

Refracted light

Glass plate

Bottom reflection surface

Transmitted light

# Summary

We can only understand materials by first understanding their properties and how the properties limit applications. Therefore, this introductory chapter has been limited to those properties we must be familiar with in order to understand polymers, ceramics, and composite materials. The most frequently specified properties are the mechanical properties, particularly strength. Toughness reflects energy absorption in breaking and is usually measured by impact testing. Perhaps the most important concept for structural applications is the effect of stress concentration, which causes failure to occur at otherwise reasonable stress. Other properties that we must be concerned with are electrical, thermal, and optical properties.

# Terms to Remember

| | |
|---|---|
| compression | fatigue |
| dielectric strength | fracture toughness, $K_{1c}$ |
| ductility | hardness |
| elastic deformation | impact |
| elastic modulus | laser |
| electrical properties | mechanical properties |
| elongate | optical properties |

physical properties                stress concentration

plastic deformation                tension

resistance                         thermal shock

resistivity                        toughness

strain                             ultimate tensile strength

stress                             viscoelasticity

# Problems

1.  Using the descriptions in Figure 1.3, list two examples of each of the following:
    a.  a brittle material
    b.  a soft, weak material
    c.  a hard, tough material

2.  A 12-in. long rectangular $Al_2O_3$ bar, 0.125 in. $\times$ 0.50 in. in cross-section, is subjected to a 750-lb tensile force. Calculate the consequent stress and strain in both British and SI units. (The elastic modulus, $E$, of $Al_2O_3$ is $55 \times 10^6$ psi.)

3.  Describe how to determine tensile strength and elastic modulus of a brittle ceramic material.

4.  A load of 5000 lb is applied on top of a brick that is 4.5 in. $\times$ 9 in. in cross-section and 3 in. high. What kind of stress results? What is the new height of the brick if the modulus of elasticity, $E$, is $1.3 \times 10^6$ psi?

5.  Synthetic ropes are sometimes made from polypropylene fibers that are strong but extend elastically much more than other rope fibers. What dangers are posed when a polypropylene rope fails?

6.  What considerations should be made when selecting the polymer material to be used for the toolbox pictured in Figure 1.7?

7.  Explain in your own words what thermal shock is. Name two specific materials that are particularly susceptible to thermal shock and describe what factors other than material selection might increase or decrease this susceptibility to thermal shock.

8.  Describe in your own words what toughness is.

9.  List four materials that display good thermal and electrical conductivity and four that have good thermal and electrical resistance.

10. Explain why glass and metal that are to be joined together must have similar coefficients of expansion.

# 2

# *Polymer Materials*

Polymers are commonly called plastics even though plastics make up one class of polymers that can be shaped. Historically, polymers are the most recent materials to be embraced by mankind. Perhaps this sense of time is best characterized by the now classic movie from the sixties, *The Graduate*, in which Dustin Hoffman's character was advised to go into plastics because that's where the future would be. Actually, there are natural polymers, such as wood and cotton, that have long been known, but the vast number of today's polymeric materials are man-made.

The term **polymer** comes from the Greek language and means many (*poly*) units (*mers*). Polymers are organic materials made up of long-chain molecules containing carbon and hydrogen in combination with many other elements. The term **plastic**, however, is correct only when referring to organic long-chain molecular solid materials that were at one time liquid and are capable of being formed into shape by heat and pressure. Another term, **elastomer**, describes certain polymers that can be stretched more than 200% under a tensile load and yet recover their original shape when the load is removed.

Most polymers are synthetic, derived to a large extent from petroleum. Others are derived from natural **resins** such as gutta-percha, rosins, tar, and other modified natural sources such as cellulose and protein. Products made from plastic materials are far-reaching, ranging from all types of packaging to medical products to insulation to materials used in the construction, communication, and transportation industries. In America alone, there are some 25,000 companies that annually produce many billions of pounds of shaped plastic parts for industrial and consumer use.

In contrast to metals and ceramics, which have been in use for many millennia, plastics are very recent material developments. One of the first known synthetic polymers was vinyl chloride, a white residue that in 1835 was extracted from ethylene dichloride in alcohol. This material was not exploited for almost a century, however. In 1843, a natural polymer, **gutta-percha**, became the first commercial polymeric material, used for knife handles and whips in Malaysia. Michael Faraday discovered that gutta-percha was a good electrical insulator, still a major application for today's polymers. In the United States, the first synthetic polymer to be developed was celluloid, produced by J. W. Hyatt in 1868 from pyroxylin, a nitrocellulose with low nitrogen content. Celluloid was used to replace ivory and gutta-percha in such items as billiard balls, combs, and brushes. The polymer industry first gained momentum in the beginning of the twentieth century, however, when L. H. Baekeland produced phenol formaldehyde resins, known since then by the trade name **Bakelite**.

In this chapter, we are going to learn about the nature of polymers and how they are formed as long-chain, high molecular weight structures. Some of these long-chain molecules are linear, or **aliphatic**, polymers and some are **aromatic** polymers that contain rigid hydrocarbon rings. Some are also three-dimensional structures formed by the branching of linear chains. We can also classify polymers by other methods, such as their source, type of polymerization reactions, morphology (i.e., crystal structure), or by end-use properties, such as light penetration, reactions to heat, and so on. In Chapter 3, we will look more closely at specific polymer materials and their processing.

# 2.1   Polymer Materials

The major bonds that hold atoms together are *covalent*, where electrons are shared by adjacent atoms, *ionic*, where electrostatic attraction occurs when electrons are donated or accepted to provide complete electron shells, and *metallic*, where essentially mobile covalent bonds allow free electrons to be shared among many atoms. There are also secondary bonds that are relatively weak by comparison. These are seldom mentioned when studying metals and ceramics because they are not important. In polymer materials, covalent bonds provide strong, rigid bonding of atoms within a long-chain molecule while the secondary bonds provide the attraction forces between long-chain molecules.

The simplest polymer is **polyethylene**. The ethylene **monomer** has a strong double covalent bond that must be broken to form the **mer**, or repeating unit, of the

**Figure 2.1**
The ethylene monomer and repeating unit,
or mer, of polyethylene

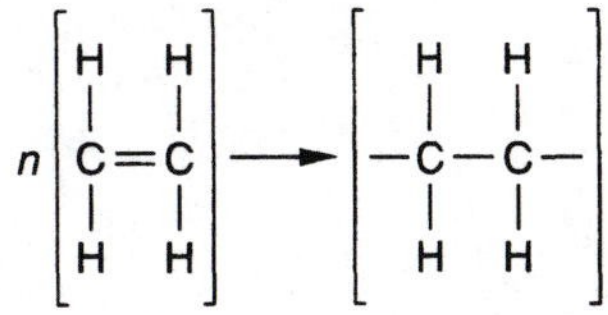

Ethylene    Polyethylene

long chain that can then be *polymerized* to form the polyethylene, as shown in Figure 2.1. We can determine the molecular weight of a polymer by multiplying the molecular weight of the mer by the degree of polymerization, $n$. Most commercial polymers have values of $n$ in the range of 25,000 to 100,000. For polyethylene the molecular weight of the monomer, $C_2H_4$, is $2 \times 14 + 4 \times 1 = 32$ and the molecular weights of the polymer would be in the range $8 \times 10^5$ (25,000 × 32) to $3.2 \times 10^6$ (100,000 × 32).

The properties of polymers are affected by the degree of polymerization. In general, low degrees of polymerization are achieved for **coatings** or surface finishes whereas high degrees of polymerization are necessary for polymers that are shaped by extrusion or injection-molding processes. At this time, though, we are not looking at processing, but at the different polymer chemistries. In polymers we have the ability to alter properties by substituting a radical, that is, a group of atoms, into the long-chain molecule. By doing so, we form materials with significantly different characteristics, characteristics that can be tailored to specific needs in many instances.

In order to understand the role of composition in determining the composition-structure-property relationship, we must be able to distinguish certain characteristics in order to classify the numerous polymer combinations. This is best done by looking at the two common methods of polymerization that can convert the monomers to a polymer. These processes are addition polymerization, which accounts for the vast majority of all plastics that are produced each year, and condensation polymerization. In addition polymerization, the repetitive mers are added to the long molecular chain with no by-product generated during the polymerization. Polyethylene is an addition polymer. **Condensation** polymers are the products of two starting materials that react and form the polymer and a by-product. For example, Baekeland's phenol formaldehyde resins react under heat and pressure to form phenol formaldehyde (Bakelite) and water, as shown in Figure 2.2.

Another method of distinguishing the many polymers that are commercially available is by the extent of cross-linking between the long-chain molecules. When there is no cross-linking, polymers can be reshaped by heating to elevated temperatures; we call these polymers **thermoplastics**. On the other hand, when extensive cross-linking does occur, the polymer is rigid and cannot be reshaped when heated; we call such polymers **thermosetting** plastics. All thermosetting resins, such as the phenolics and epoxies, are condensation-type polymers, whereas only a few engineering thermoplastics, such as nylons, polyesters, and polycarbonates, are condensation-type polymers. Addition-type polymers, including **polyethylene, polypropylene, polystyrene, polyvinyl chloride**, and many others, are thermoplastics.

**Figure 2.2**
Condensation polymerization of phenol formaldehyde

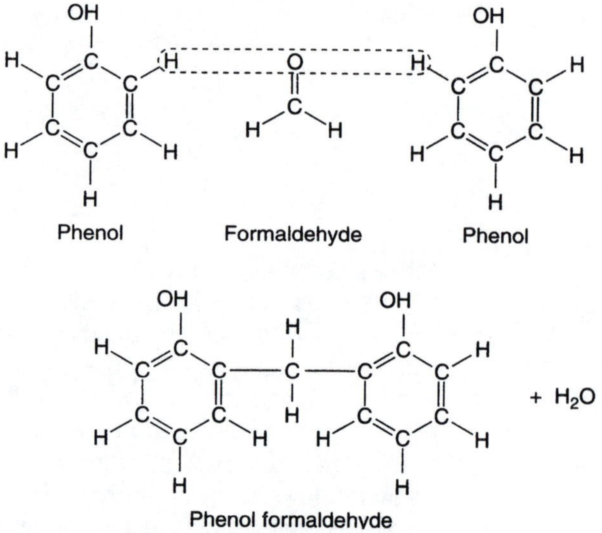

Phenol  Formaldehyde  Phenol

Phenol formaldehyde

In polyethylene we can substitute a side group, $R_1$, for hydrogen and form the *vinyl* linkage, as illustrated in Figure 2.3. Common polymers based on this vinyl linkage are identified in Figure 2.4.

# 2.2  *Structure of Polymers*

The nature of long-chain, high molecular weight polymers leads to diverse structures of the molecules. *Linear* structures have long chains (either aliphatic or aromatic) that are only loosely bonded together by secondary bonds. *Branched* structures, in which the branches are bonded covalently to the main chain, provide additional mechanical interlocking of the long chains. *Cross-linking* is the bonding of linear chains together by low molecular weight compounds. In some cases, we can produce these structures in the same material, depending on how we polymerize it. In polyethylene, for example, we can make linear polyethylene, branched polyethylene, or we can chemically treat or irradiate polyethylene to develop cross-linking, thus providing improved heat resistance and tensile strength.

In addition to these polymer structures, we can also modify the structures by combining more than one monomer in the same chain, thus obtaining different

**Figure 2.3**
The vinyl linkage

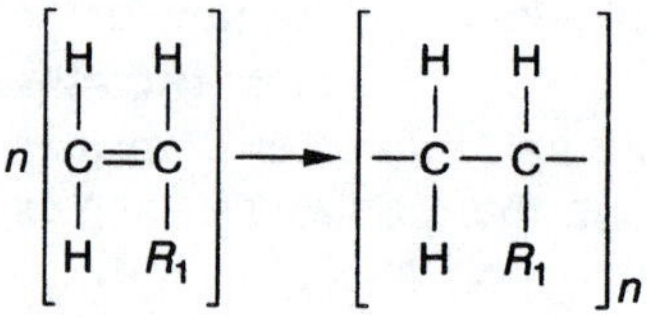

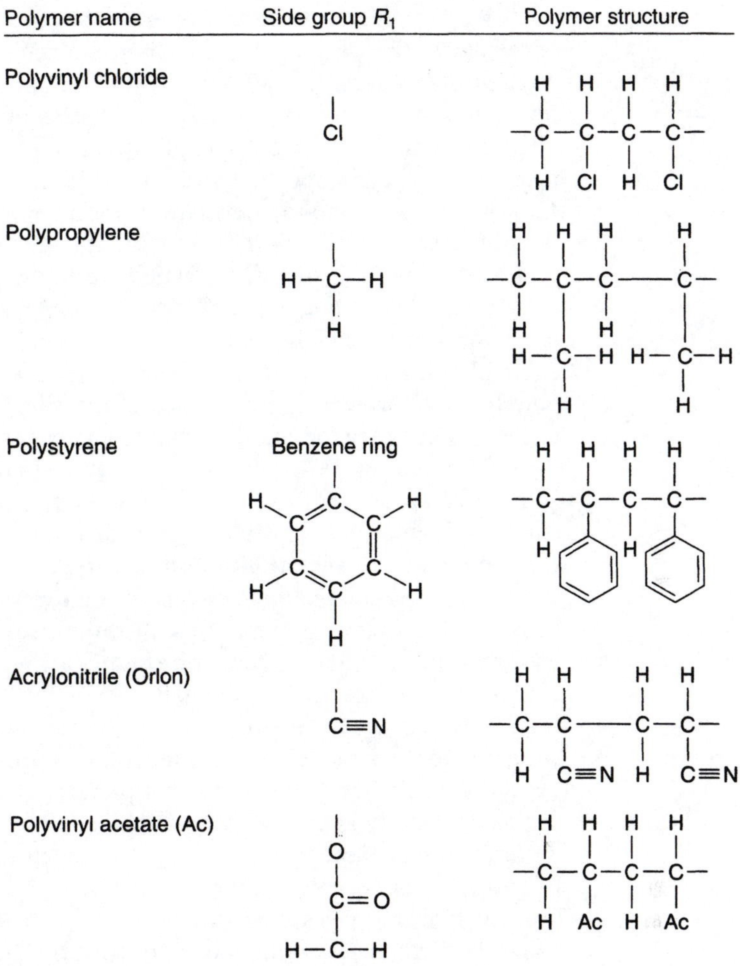

**Figure 2.4**
Polymers based on the
vinyl linkage

| Polymer name | Side group $R_1$ | Polymer structure |
|---|---|---|
| Polyvinyl chloride | | |
| Polypropylene | | |
| Polystyrene | Benzene ring | |
| Acrylonitrile (Orlon) | | |
| Polyvinyl acetate (Ac) | | |

properties. We call a polymer that contains two monomers a **copolymer** (e.g., styrene acrylonitrile) and one that contains three monomers a **terpolymer** (e.g., **ABS**, or **acrylonitrile-butadiene-styrene**). These combinations of monomers can be *alternating* (–ABABABABABABAB–), *random* (–ABAABABBBABABBAA–), or *block* configurations (–AAAAAABBBBBBAAAABBBBBAAA–). We also can form **graft copolymers**, which are similar to branched structures, with a backbone of one monomer and branches of another:

```
AAAAAAAAAAAAAAAAAAAAAAAAAAAAAAAAAAAAAAAAAAAA
    B     B     B     B     B     B     B
    B     B     B     B     B     B     B
    B     B     B     B     B     B     B
    B     B     B     B     B     B     B
```

We normally consider polymers to be amorphous, or noncrystalline, materials, but order to the packing of long-chain molecules actually provides a wide range of crystallinity. In fact, crystallinity is a function of the polymer chemistry, the inherent polymer structure, and mechanical processing. Some of the properties that are affected by the degree of crystallinity are opacity, mechanical strength, chemical resistance, surface smoothness, and permeability to gases.

Most polymers are transparent when amorphous, but become opaque as crystallinity develops. One of the factors that can contribute to the orderly packing into crystalline arrays is diffusion, which we know to be affected by atomic size and vacant sites. Using the vinyl link as an example, polyethylene is made only of carbon and hydrogen atoms, the smallest of all atoms. These diffuse readily and polyethylene with a high degree of crystallinity is readily achieved. For polyvinyl chloride or polystyrene, however, the large size of the chlorine atom or the benzene ring limits diffusion necessary for chain proximity and crystallization.

Proximity of molecular chains is also affected by branching, which interferes with crystallinity. The proximity can be enhanced, however, if the geometry of the chains is regular. We call the degree of regularity *stereoisomerism* and it can take three basic forms, shown in Figure 2.5. **Isotactic** and **syndiotactic** arrangements of the side groups can facilitate chains fitting together more readily, promoting crystallization. In order to prevent crystallization, careful control of the polymerization process can form the random, or **atactic**, arrangement, which will not permit such close packing of the molecular chains.

We must keep in mind that polymer chains are very long and are never stretched to their full length. It is much more appropriate to think of the structure as twisted or entangled chains. In this picture, the "crystals" are extremely fine and chains actually pass through an amorphous region between adjacent crystals. Thus we never can achieve 100% crystallinity. Because the chains are tangled and twisted, however, we can draw or deform them, thus aligning the chains in the working direction. Such techniques are used to develop the highest degree of crystallinity possible, about 98% in the case of high-density polyethylene.

# 2.3   Additions to Polymers

In addition to the varied compositions and structures of the long-chain polymer molecules, we can alter many properties by the addition of both organic and inorganic materials. For example, we can add agents that cause foaming or expansion during processing and we can reinforce polymers with many fibrous and other filler materials. Some of the reasons we add materials that alter the polymers are to

improve mechanical, physical, or chemical properties

improve processing and control of processing

add color

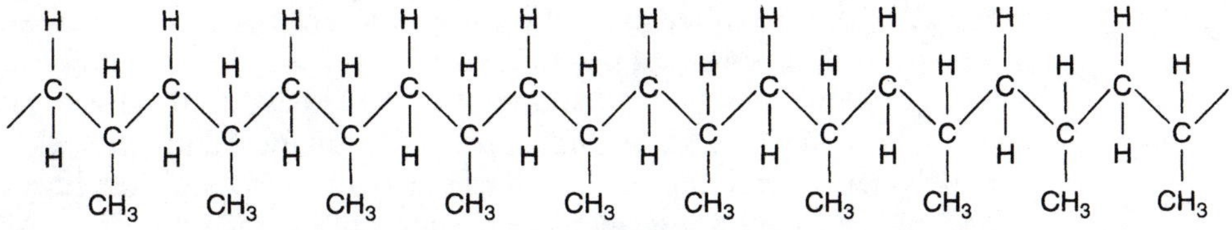

**Figure 2.5**
Stereoisomers of polypropylene: (a) isotactic (same $CH_3$ arrangement), (b) syndiotactic (regular $CH_3$ arrangement), (c) atactic (random $CH_3$ arrangement)

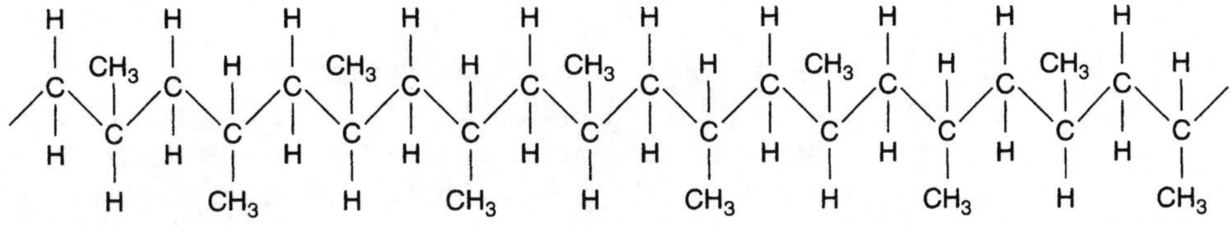

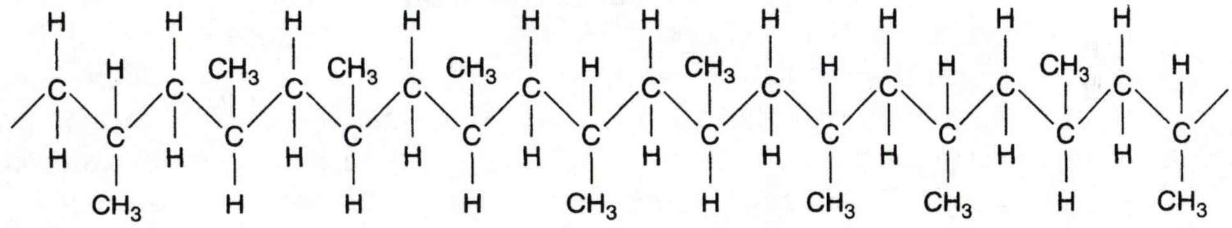

> reduce degradation in service
> reduce cost

Let's look at these in some detail.

## 2.3.1 Improved Properties

Perhaps the most important properties of all materials are the mechanical properties, particularly strength and ductility. Polymers in general are not as strong as metals, alloys, and ceramics and consequently must be reinforced with additives for applications that require high strength. The composites that are formed will be described in detail in Chapter 8, but at this time, we need only be concerned with the broad characteristics. Composites are conveniently divided into three groups—fibrous, particulate, and laminar. The properties are, of course, dependent on direction because of the nature of the combination, but they also depend on the properties of each material, the size, shape, and amount of the additive, and the bond between additive and matrix.

We also can alter properties in a different manner to increase flexibility, not strength. Additives called **plasticizers** not only increase flexibility, but improve processing by lowering the melt temperature and the viscosity. These plasticizers can be organic chemicals, such as phthalates, sulfonamides, and adipates, or they can also be polymeric. They can be used for linear, branched, or cross-linked polymers, and they find extensive applications in coatings, films, and extrusions. Alloying polymers with elastomers to form the polyblends can also provide the same benefits in improved flexibility, but the purpose for these additives usually is increased impact resistance.

Polymers are generally excellent electrical insulators and are applied widely in the electrical and electronic industries. However, they remain susceptible to problems associated with static electricity and electrostatic discharge. By incorporating antistatic additives such as polyethylene glycol esters or amines, polymer surfaces become more conductive and therefore less susceptible to accumulated static electric charge because the additives attract moisture from the air.

Additives that are flame retardants are important because all polymers will ignite and burn under certain conditions. Chemicals that make it difficult to ignite a polymer as well as slow the combustion process can be added. However, a complicating factor is that particular additives may be effective for one polymer but ineffective for others. Both organic halogenated hydrocarbons and inorganic salts are used. These additives act by promoting charring, thus forming a glassy insulating layer that cools or interferes with the combustion process (called *chain-breaking*).

# 2.3.2   *Improved Processing or Process Control*

Many additives provide control of the polymerization or shaping of polymers. For example, thermosetting resins might have *inhibitors* for long-term storage before polymerization or *catalysts* that initiate polymerization. Sometimes *accelerators*, which speed the curing process in the presence of a catalyst, are also added. Shaping processes are benefited by addition of lubricants that reduce friction between the workpiece and equipment, reduce internal friction, and prevent sticking. *Solvent chemicals* and *water* are also common processing aids; they dilute the polymer to make application easier, then evaporate to leave a film coating.

Additives are also necessary to produce low-density cellular foam products. *Foaming* or *blowing agents* can be chemical, where decomposition occurs at specific temperatures and releases gases that form the cellular foam structure, or they can be physical, where gases are dissolved in the melt because of the process pressure. The gases evaporate when pressure is released as the melt leaves the die and expand the polymer.

The strength of fiber-reinforced composites, as we will learn in Chapter 7, depends on transfer of the stress from the weak matrix to the strong fiber. In order to successfully transfer the stress, there must be a good bond between the two different materials. During processing, surface treatments that improve these bonds are aided by the addition of *coupling agents*.

## 2.3.3 Color Additives

Polymers differ from the other materials we have studied because they can be colored, a feature that makes them particularly attractive for commercial applications. The most common coloring method utilizes the addition of pigments, which are dispersed throughout the polymer. The vast number of pigments available is easy to determine by simply reading the ingredients on labels of polymer paint coatings. Some pigments provide special effects, such as metallic finishes that incorporate metal flakes in lacquers and fluorescent pigments used in safety signs and reflective clothing.

## 2.3.4 Reduction of In-Service Degradation

Many polymers are subject to the degradation of properties and appearance during their normal lifetime. For example, colors can be altered by sunlight and weathering, which can lead to fading and/or discoloration. Polypropylene and polyethylene are susceptible to oxidation, which embrittles the material. Surface cracks formed in a high-density polyethylene beer crate weathered for 9 years are shown in the scanning electron micrograph of Figure 2.6. Other polymers, such as polyvinyl chloride, polystyrene, polyolefins, ABS terpolymers, and polyurethanes, are affected adversely by ultraviolet radiation. Even biological breakdown can occur, such as in the surface of the polyvinyl chloride shown in Figure 2.7 that was exposed in a compost heap.

**Figure 2.6**
Surface of high-density polyethylene crate after 9 years of use and weathering (120×) (From L. Engel, H. Klingele, G. W. Ehrenstein, and H. Schaper, *An Atlas of Polymer Damage,* Prentice-Hall, Inc., 1981.)

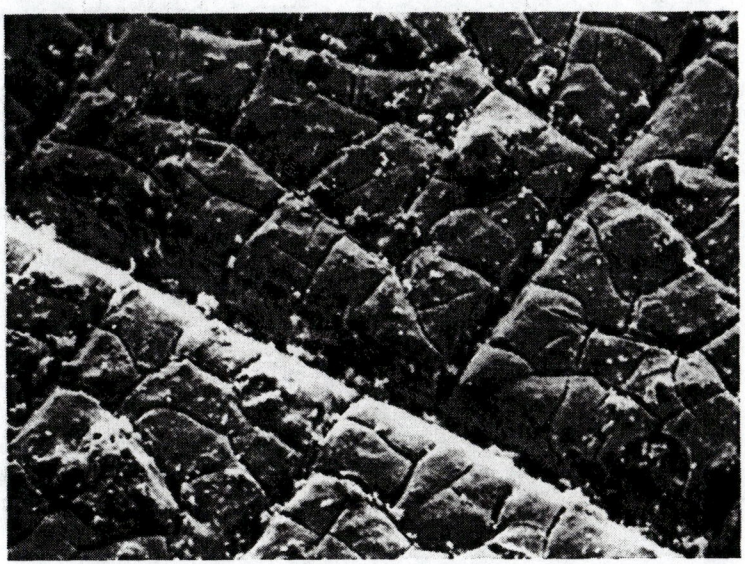

**Figure 2.7**
Pits and scars in polyvinyl
chloride surface caused by
microorganisms (1000×)
(From L. Engel, H. Klingele,
G. W. Ehrenstein, and H.
Schaper, *An Atlas of Polymer
Damage,* Prentice-Hall, Inc.,
1981.)

In order to prevent degradation, liquid or powder *stabilizers* are added to the polymer. These stabilizers include antioxidants, such as alkyl phosphites that make the polymer chain inert, and ultraviolet stabilizers, such as hindered amines that absorb large amounts of energy and form a chelate that then releases the energy at a lower level that does not damage the polymer.

# 2.4   General Applications of Polymers

The largest volumes of polymers are used in the packaging and construction industries, but many other substantial markets also exist, including electrical and electronics, transportation, appliances, housewares, furniture, and toys. Polymers have found extensive applications across many of these industries. In fact, there are three polymer types that have dominated the market because of their extensive applications across-the-board and because of their inexpensive price. These are the **polyolefins**, such as high density polyethylene, the *vinyls*, such as polyvinyl chloride, and the *styrenes*, such as polystyrene.

Many **films** are widely used as packaging for food products, produce, and clothing, as backing for adhesives, and as displays for sale merchandise. They are also used for fillers in packaging in the form of bubble wrap. Film materials include polyethylene, polyvinyl chloride, polyvinylidene chloride, polyurethane, and many others. Expanded rigid polystyrene foams find many uses as lightweight fillers and shock absorbers in shipping, many of which are molded to protect sensitive contents of packages. Even plastic fibers have found extensive use in packaging; polyolefin fiber mailing envelopes have replaced many weaker paper products.

In the construction industry, many of these same polymers are used extensively. The polyolefin fiber materials are used as housewrap, which effectively reduces the heat losses caused by drafts, and rigid polystyrene foam is used for thermal insulation. Acetals are used for plumbing fixtures and rigid polyvinyl chloride (PVC) is used extensively for plumbing connections and as siding for houses. With plasticizers added, PVC is used for electrical wiring insulation. Silicones are used for caulking applications and latex polymers are used in painting. Even modern countertops are made from thermosetting laminates containing mica for heat resistance. Other examples of polymer uses in the construction industry abound and increase almost daily. Even recycled polymers are being introduced in outdoor landscaping applications where resistance to rotting is necessary and as nailable two-by-fours for framing applications.

There are many diverse polymer usages in other fields as well. In electronics, for example, phenolics find applications as chip carriers, epoxies as printed wiring boards, polyurethanes as conformal coatings, polyamides as coatings for multilayer laminates, and polycarbonates as connector blocks. In transportation, polymers are used in auto bodies, seat coverings, battery cases, dashboards, steering wheels, and many more parts. In some airplanes, polymers and composites account for as much as 30% of the material parts.

Other industries in which polymers find extensive applications include toys, sports equipment, clothing, furniture, housewares, agricultural equipment, and appliances. Despite the versatility and diverse uses of polymers, however, we must remember that polymers are not cure-alls and that good engineering concepts must always be practiced. Case Study 2.1 is an unfortunate example of where these engineering principles were overlooked.

---

## Case Study 2.1

### Stress Concentration Again!

It was a beautiful, sunny Sunday morning in July. John was enjoying his first speedboat ride off Cape Ann, Massachusetts. The thrill of barely touching the waves at nearly 80 mph was far greater than he had anticipated, but that thrill was short-lived. The boat suddenly veered sharply to the right, throwing John into the ocean, but not before his leg was broken by contact with the low framework of the passenger compartment.

John was seriously injured by an event that need not have happened. Accident reconstruction showed that the steering loss and sudden right turn resulted from failure of a molded nylon clamp holding one cable of the boat's dual rack-and-pinion steering control. With the clamp broken and the cable loose, water forces acting on one side of the outboard motor could not be controlled.

Figure 2.8 shows the broken clamp; fracture occurred through holes where the clamp was affixed with bolts to the housing for the rack-and-pinion gear and through the reduced section where the cable was seated. It is obvious that this failure was caused by poor design where the stress concentration permitted the local stress to exceed the breaking strength of the nylon. Failure could be prevented only by redesign or substitution of a stronger material than the molded nylon.

From "Materials and Product Safety," *Professional Safety,* Vol. 36, No. 4, 177, 1991.

# 2.5  Polymer Classifications

There are numerous ways to classify polymers for closer examination. We have already looked at one method—the chemistry of the molecular chain. We can also compare them by their most common form, such as molded or fibrous, by their applications, such as engineering or consumer, by their properties, such as elastomers or rigid polymers, by their processing characteristics, such as thermoplastic or thermoset, and even by cost. None of these classifications is totally satisfactory, but we will use processing only because the actual processing of thermoplastics and thermosets is drastically different, as we will learn in Chapter 3. In the following sections, we will examine the major features, including properties, forms, compositions, and applications, of the more popular industrial polymers.

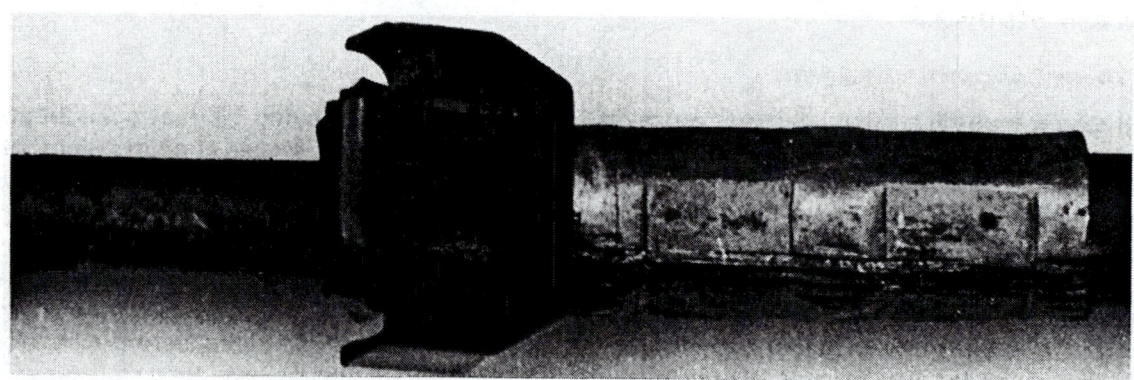

**Figure 2.8**
Portion of fractured nylon clamp remaining with speedboat steering cable.

# 2.5.1 Thermoplastic Polymers

If we consider consumption as a critical criterion, then the three most important thermoplastics are the polyolefins polyethylene and polypropylene, polyvinyl chloride, and polystyrene.

*Polyethylene* is inexpensive, has a range of properties, is resistant to moisture and chemicals, is a good electrical insulator, and can be processed by all conventional thermoprocessing techniques. However, it is oxidizable, is difficult to bond to, and cracks easily under stress. We frequently differentiate the polyethylenes by density. Low-density polyethylene (LDPE) has a specific gravity of 0.91, is branched, flexible, and used for applications such as bottles. LDPE should not be confused with very low-density polyethylene (VLDPE) or linear low-density polyethylene (LLDPE), which are actually copolymers. LDPE, which has tensile strengths of 600–2300 psi, is used as film and flexible tubing, whereas VLDPE is used for applications such as shrink-wrap and squeeze tubes and LLDPE is used for many packaging applications, such as dry cleaners' garment bags. High-density polyethylene (HDPE), with specific gravity of 0.95 and strengths of 3100–5500 psi, has no branching and is more crystalline. It is used for containers for gasoline and for industrial pipe.

*Polypropylene* has the vinyl linkage with the $R_1$ group being the $CH_3$ group. It is cheaper than polyethylene because propylene gas is cheaper than ethylene and it is lighter, with specific gravity of 0.90. Polypropylene is also more rigid than polyethylene and can be used at higher temperatures. Commercial polypropylene is highly crystalline, formed from isotactic molecular chains. It is used for chair seats and backs in office furniture, for dishwasher parts, and for containers with integral hinges and covers.

Both polyethylene and polypropylene can be produced in fibrous form and can be expanded to make foam products. Polyethylene fibers are used for disposable clothing and for housewrap, which reduces air drafts and moisture transmission but allows the house to breathe. Polypropylene fibers are used for strong, water- and mildew-resistant ropes.

*Polystyrene* also has the vinyl linkage with the $R_1$ group being the benzene ring. It is atactic, amorphous, and has high strength, 5,000–12,000 psi, and good thermal and electrical resistance. Common applications are in TV cabinets, wall tiles, disposable dishes, furniture, molded parts, and containers. About a third of all the polystyrene produced is expanded cellular foam familiar perhaps to all of us in the form of hot drink cups.

Styrene has also found many applications when copolymerized with acrylonitrile ($CH_2=CHCHCN$). **Styrene acrylonitrile**, known as **SAN**, has improved chemical resistance and impact strength. Some common products made of SAN are telephone cases, food processor bowls, and medical syringes. Highest impact resistance is achieved by a terpolymer of polystyrene acrylonitrile and butadiene ($CH_2=CH–CH=CH_2$) known as ABS. These polymers are used for automotive parts, housings for tools and appliances, luggage, and safety hard hats.

*Polyvinyl chloride* is the last of the three largest production polymers. As we already know, PVC has the vinyl linkage with the $R_1$ group simply the chlorine atom. The volume of PVC consumed annually is attributable to its flexibility in properties, achieved by the addition of plasticizers. PVC's largest applications are as films and sheet materials, which are widely used as collapsible containers, wallpaper coverings, auto seat covers, clothing, and leather substitutes. Rigid PVC is used for construction siding and gutters and plumbing pipes, and plasticized PVC also is used in flooring, for flexible hoses, and as insulation for electrical wiring. PVC can be foamed and laminated, a technique used in making shoes. However, when PVC is used in applications where heating occurs, chlorine gas can be released, forming hydrochloric acid that chemically attacks metals in the vicinity.

## Case Study 2.2

### Failure of a Polyvinyl Chloride Pipe

The East Campus of the University of Massachusetts Lowell is located several miles from the main campus. Recently, a new maintenance facility was approved and built adjacent to some protected wetlands. Electrical service to the building was buried 2 ft below the surface and high-voltage, shielded cable was inserted through a 1½-in. diameter Schedule 40 rigid PVC pipe. Because of the distance to the building; however, the cable had to be spliced. When making this type of joint, the metal shielding as well as the cable must be properly joined. The actual joint of this shielding was inadequate; however, it did not cause any problems until after a bitter New England cold spell. The power was lost to the entire building because the shielding joint had corroded through and caused arcing and subsequent separation of the spliced main cable.

The corrosion was caused by water that had entered the PVC pipe through a puncture near the spliced joint, found only when the PVC pipe was excavated. The cracked wall of the pipe is shown in Figure 2.9. It was clear that after the original excavation, although done according to specifications for backfill and compaction, a 1/2-in. stone was left resting against the pipe. Because of the adjacent wetlands, moisture had penetrated through the sand and gravel around the pipe, then froze and expanded, forcing the stone through the rigid PVC pipe wall. When thawing occurred, water had no barrier to entry into the pipe.

Beyond the top three thermoplastic polymers are a large variety of alternative compositions, but we will only look at those that have interest in industrial applications. These are the polyacetals, polyamides, acrylics, and polycarbonates.

**Polyacetals** are linear, highly crystalline polymers selected frequently for their ease in moldability, superior fatigue resistance, stiffness, and resistance to water. The proper name for this polymer is polyoxymethylene (POM), but we refer to it by the generic term *acetal* by convention. Industrial applications are found in automobile carburetors and door handles, in videocassette parts, in tool handles, and in plumbing parts.

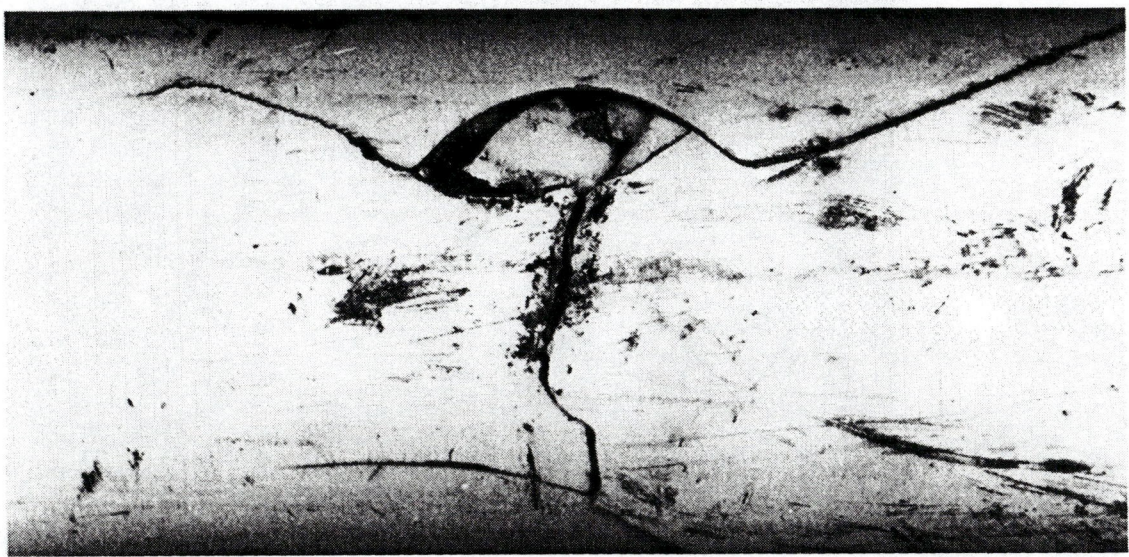

**Figure 2.9**
Fracture and cracks caused by a stone pressed against a PVC pipe by freezing water
in the soil

## Case Study 2.3

### Stress Concentration and Fatigue

Flexible hoses are a convenient means for connecting plumbing fixtures, eliminating the need for cutting copper tubing and brazing it into place, often in cramped quarters. When Loring Corporation was remodeling their offices, flexible fittings were installed for the water supply to a toilet in a remote rest room. Directions for installation of the polyacetal nut were to tighten one half turn beyond hand tight and *not* to overtighten. Polyacetal was selected because of its excellent fatigue resistance. Fourteen months later during a plant shutdown, the nut failed, causing significant water damage before the leak was discovered.

In conducting a failure analysis, an exemplar unit was tightened until failure. Following installation directions, the one half turn beyond hand tight corresponded to 22 in.-lb torque. Continued tightening caused failure at 120 in.-lb torque, in good agreement with the manufacturer's data. The compressed washer and fracture surface of this exemplar failed unit and the actual failed unit were examined in a scanning electron microscope. Figures 2.10 and 2.11 demonstrate that both washers were compressed to nearly the same extent. However, the fracture surfaces of the exemplar unit, Figure 2.12, and the subject unit, Figure 2.13, indicate fracture propagated in different directions.

**Figure 2.10**
Compression of exemplar washer in unit tested to failure in torsion (13.5×)

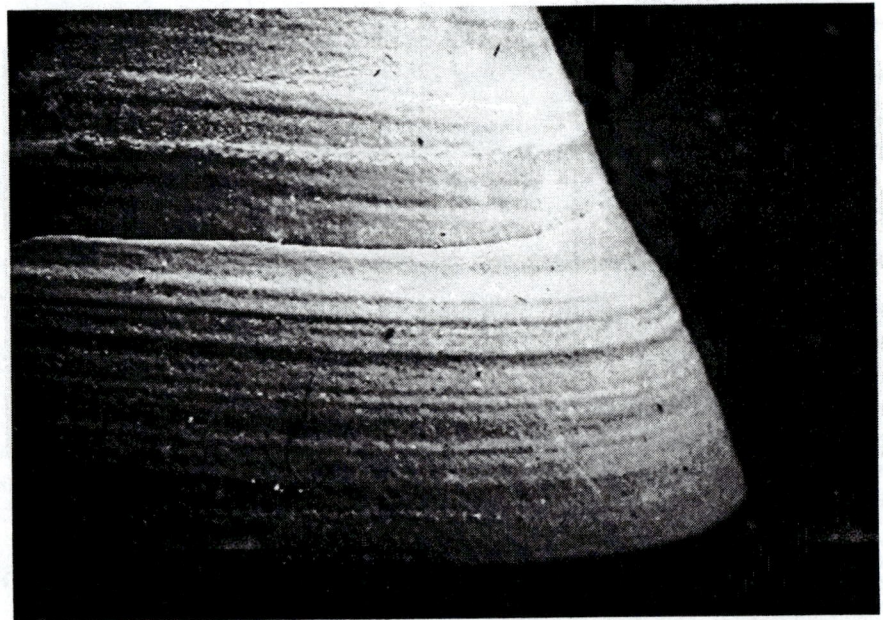

**Figure 2.11**
Compression of washer in failed unit (13.5×)

**Figure 2.12**
Fracture surface of exemplar coupling nut intentionally broken in torsion (24.8×)
(From D. A. Colling, "Case Studies—Polymers and Products Liability," ANTEC 94
Paper, Society of Plastics Engineers, Brookfield, CT.)

It was concluded that the failure occurred because of improper installation. An incipient crack formed by overtightening propagated slowly because of fatigue stresses associated with intermittent running water. The magnitude of the stresses was sufficient to cause fatigue failure only because of the stress concentration created by the geometry of the incipient crack, that is, a long crack length with small tip radius.

**Polyesters** derive their name from the combination of *poly*merization and *ester*ification. In esterification, an organic acid is combined with an alcohol to form an ester and water. Unsaturated polyester resins are thermosetting and are often used as the matrix for glass fiber reinforced structures (fiberglass). Saturated polyesters, such as **polyethylene terephthalate (PET)**, are formed by the combination of terephthalic acid and ethylene glycol. These are probably known better by the trade names Mylar in film form and Dacron in fiber form. PET is highly crystalline and oriented for most applications, has a high glass transition temperature with good mechanical properties to temperatures as high as 175°C. It can be produced as film that is used for decorative purposes, such as metallized balloons, for tapes of all kinds used in photography and videotaping, and in backing for adhesives. In fiber

**Figure 2.13**
Fracture surface of failed coupling nut (24×)
(From D. A. Colling, "Case Studies—Polymers and Products Liability," ANTEC 94
Paper, Society of Plastics Engineers, Brookfield, CT.)

form, polyesters are moisture resistant and crease resistant; blended with cotton, they are the standard for wash-and-wear garments.

**Polyamides** are better known by their generic name **nylon**. Nylons are described by a numbering system based upon the number of carbon atoms in the monomers. They are highly crystalline, tough, and strong (9,000–12,000 psi tensile strength), and have a low coefficient of friction. Easily machinable, molded forms are used as bearings, gears, furniture casters, plumbing connections, and in many other applications. Spun into fibers, nylon is used for tire cord, umbrellas, clothing, carpeting, camping equipment, and many other durable products. Many nylons are even used as hot-melt adhesives for book bindings.

Aromatic polyamide (aramid) fibers, better known by the DuPont trade name **Kevlar**, have nearly twice the stiffness and half the density of glass fibers. Kevlar is used in bullet-resistant clothing and in composite forms as armor, military helmets, aircraft structures, and boat hulls.

Thermoplastic **acrylics** include acrylate and methacrylate, the most important of which are **polymethyl methacrylate** (PMMA) and **polyacrylonitrile** (PAN). These polymers, whose monomer structure appears in Figure 2.14, have good optical clar-

**Figure 2.14**
Acrylic mers

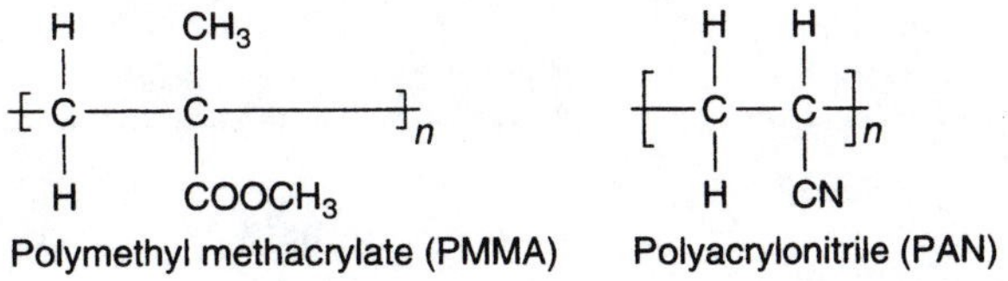

Polymethyl methacrylate (PMMA)          Polyacrylonitrile (PAN)

ity, can be made in many colors, and have good rigidity and impact strength. Such qualities make PMMA widely used as a substitute for glass windows, contact lenses, and auto lenses. PMMA can be alloyed with PVC to produce tough, durable sheet that is formed into advertising signs.

PAN fibers, such as **Orlon**, have high strength, stiffness, toughness, and resilience and are typically used for clothing. However, they also are converted at high temperatures to carbon fibers that are used in fiber composites (the conversion is described in Chapter 8).

**Polycarbonates** (PC) are characterized by toughness, resistance to high temperatures, and good dimensional stability. Their structure, given in Figure 2.15, has both the benzene ring and the –OCOO– carbonate unit that repeats and provides the unique characteristics. Applications based on these properties include the compact disks cherished by music lovers, housings for home and construction power tools, hair dryers, and toasters, and even safety hard hats and hot beverage dispensers.

Table 2.1 compares the properties of selected polymers from each of the thermoplastics we have been studying. The only major thermoplastic material missing from this table is PVC, which has not been included because of the extensive use of plasticizers in it.

# 2.5.2  *Thermosetting Polymers*

The first commercial thermosetting polymer to find extensive use was Bakelite, the trade name for the phenolic thermosetting material developed in 1909 by Dr. Baekeland. Although thermoplastic polymers are used much more extensively today, production of phenolics remains only behind the polyolefins, polyvinyl chlorides and styrenes. Thermosets are characterized by cross-linking of the molecular chains, providing rigidity. Of course, once polymerized, the materials cannot be reshaped because of this irreversible reaction that forms the three-dimensional structure. Cross-linking is accomplished by heat or heat and pressure in some materials, such as the phenolics, and by chemical reaction in others, such as the epoxies.

**Figure 2.15**
The polycarbonate mer

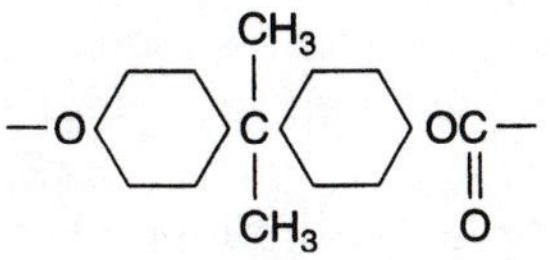

**Table 2.1**
Properties of selected thermoplastic polymers

| Polymer | Density (g/cc) | UTS (ksi) | Izod impact (ft-lb/in.) | Maximum temperature (°F) |
|---|---|---|---|---|
| LDPE | .91 | 4.0 | n/a | 104 |
| HDPE | .95 | 5.5 | 2.0 | 176 |
| Polystyrene | 1.04 | 6.0 | .4 | 167 |
| SAN | 1.07 | 8.7 | .6 | 185 |
| ABS | 1.03 | 5.0 | 6.0 | 176 |
| Polyacetal | 1.42 | 10.0 | 1.4 | 195 |
| Polyester (PET) | 1.37 | 10.4 | .8 | 175 |
| Polyamide (6,6 nylon) | 1.14 | 12.0 | 2.0 | 150 |
| Acrylic (PMMA) | 1.18 | 9.5 | .4 | 167 |
| Polycarbonate | 1.20 | 9.0 | 14.0 | 250 |

**Phenolics** are produced by condensation polymerization of phenol and formaldehyde to form an *A-stage* resin. There are both one-stage resins where an excess of formaldehyde and an alkaline catalyst are reacted to form *resol*, which is insoluble in common solvents, and two-stage resins where a deficiency of formaldehyde forms a low molecular weight *novalac*. Fillers, colorants, lubricants, and, in the case of novalac, a chemical that is capable of causing cross-linking, are added to complete this A-stage resin. There are three roles that the **fillers** can play: (1) extenders such as wood flour and mica are inexpensive and reduce cost when added in large quantities; (2) functional fillers are also inexpensive, but are added to improve such properties as stiffness, impact resistance, or shrinkage; (3) reinforcement fillers in the form of glass, graphite, and polymer fibers, although expensive, are added to increase mechanical properties. Blending these fillers and the novalac on heated rolls forms the *B-stage* resin, which is insoluble in organic solvents. The one-stage phenolics and the B-stage resin are fused under heat and pressure, converting them into the final product, called *resite* or *C-stage*, a completely cross-linked thermoset.

Phenolics generally have high strength because of fiber reinforcement. They are inexpensive, have excellent electrical resistance, can be used at temperatures up to 400°F, and are self-extinguishing. However, their dark color is not attractive and limits their use for many applications.

Formaldehyde can also react with urea and melamine to form two commercially important amino polymers. As with phenolics, they are polymerized by condensation reaction. The urea-formaldehyde molding compounds have many industrial applications, such as electrical circuit breakers and switch parts. The largest use for **melamine formaldehyde** products is for dinnerware, but there are also many construction applications where the melamine resins are used for laminates and exterior-grade plywood. The laminates are probably more recognizable by the trade name Formica, in which filled melamine surface layers are pressed onto a base material made of kraft paper impregnated with phenolic resin.

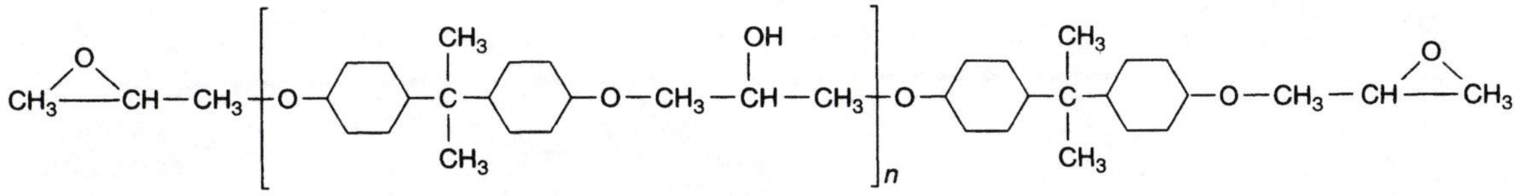

**Figure 2.16**
Epoxy resin formed by combining epichlorohydrin, bisphenol-A, and a catalyst

Polyesters were discussed under thermoplastic materials, where we learned that organic acids can react with alcohol to form an ester plus water. Unsaturated polyesters are formed by the reaction of alcohols with two OH groups (a diol) and an acid with two COOH groups (a diacid). These unsaturated polyesters are linear and can be cross-linked with vinyl-type molecules to form a thermoset. The main use for these unsaturated polyesters is the production of fiberglass-reinforced polymers used in auto body sections, small boat hulls, pressure vessels, and even welding helmets.

When modified by a nonvolatile monomer such as diallyl phthalate, polyesters are known as **alkyds**. The largest application for alkyd resins is in oil-based paint coatings, but there are moldable compounds that are used for such items as housings for tools and appliances.

**Epoxies** are thermoset polymers that have extensive applications as adhesives, surface coatings, and laminates in the electrical, electronics, aerospace, and construction industries. These applications are derived from the dielectric properties of epoxies, chemical resistance and excellent adhesion. Epoxies have the advantage of a wide range of curing conditions and properties, no evolution of volatiles during the cure cycle, suitability for all processing methods, and the ability to cross-link with other materials. Most epoxies are based upon prepolymers produced from bisphenol-A and epichlorohydrin, shown combined in Figure 2.16. Hardeners, or curing agents, which include a number of amines and acid anhydrides, are added. Cross-linking is also aided by heating the resin-hardener mixture. The amount and type of hardeners will control the properties and pot life, curing temperature, and chemical resistance, and even reduce properties that are skin irritants (many workers suffer from dermatitis caused by contact with epoxies).

Epoxy laminates find many applications as replacements for unsaturated polyester laminates because of their improved chemical and fatigue resistance, combined with reduced mass that increases the strength/weight ratio. Filled epoxies are cast into many shapes that are used for special tooling fixtures and potting of electrical components. As adhesives, they are often used for high-strength bonding of critical dissimilar materials.

There are other industrial thermosetting polymers, but we will discuss them in the following section because their major uses are as elastomer materials. Before leaving thermosetting polymers, however, we can compare typical properties of each in Table 2.2.

**Table 2.2**

Typical properties of selected thermosetting polymers

| Polymer | Density (g/cc) | UTS (ksi) | Izod impact (ft-lb/in.) | Maximum temperature (°F) |
|---|---|---|---|---|
| Phenolics | | | | |
| Wood flour | 1.39 | 7.0 | .4 | 325 |
| Mica | 1.78 | 6.2 | .4 | 275 |
| Glass | 1.80 | 6.5 | .4 | 450 |
| Polyester | 1.28 | 9.5 | .3 | 250 |
| Glass | 1.80 | 14.0 | 15.0 | 325 |
| Urea-formaldehyde | | | | |
| (Cellulose) | 1.50 | 9.2 | .3 | 170 |
| Melamine-formaldehyde | | | | |
| (Cellulose) | 1.50 | 10.0 | .3 | 210 |
| (Glass) | 1.90 | 7.5 | .6 | 350 |
| Epoxy | | | | |
| Unfilled | 1.23 | 8.5 | 5.0 | 375 |
| Glass | 1.80 | 20.0 | 20.0 | 400 |

# 2.5.3   Elastomers

Elastomers, as we learned earlier, are the polymers that extend elastically more than 200% and are capable of full recovery. They include, of course, cross-linked natural **rubber** and synthetic rubbers and some thermoplastic and thermosetting polymers. Examples are polyester thermoplastic elastomers and polyurethane and silicone thermosetting elastomers. We will examine these materials separately because many of their applications, for example, coatings, depend on their elastomer characteristics. Elastomers must be amorphous and must be used above their glass transition temperature, whereas other polymers must be crystalline or be used below their glass transition temperature for dimensional stability.

Natural rubber is derived from the latex that oozes from many tropical trees, such as the *hevea* tree. A latex is a stable aqueous solution of the rubber, which is a high molecular weight isoprene polymer. The rubber monomer is $-CH_2C(CH_3)=CHCH_2-$, which is the same as that of gutta-percha, the first known polymer. Gutta-percha, however, is not an elastomer. The difference between these two materials is caused by the geometry of the long-chain molecule. When the methyl group, $CH_3$, and the hydrogen atom are on the same side of the chain, we have natural rubber, but when they are on opposite sides of the chain, we have gutta-percha, as shown in Figure 2.17. Natural rubber by itself is a weak thermoplastic material with no useful mechanical properties. In order to become an elastomer, the molecular chains must be occasionally cross-linked, as shown in Figure 2.18, but

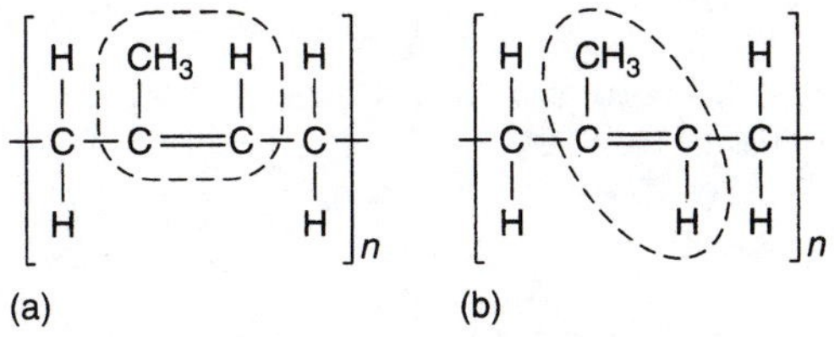

**Figure 2.17**
Geometric difference between mers of natural rubber and gutta-percha: (a) cis-1, 4 polyisoprene (natural rubber) and (b) trans-1, 4 polyisoprene (gutta-percha)

not enough to become a rigid thermoset material. This is accomplished by the process of **vulcanization**, discovered originally by Charles Goodyear in 1839. Figure 2.19 compares the stress-strain relationships of vulcanized and unvulcanized natural rubber.

Synthetic rubbers include both **styrene butadiene (SBR)**, **nitrile (NBR)**, and **neoprene** rubbers. These are copolymers, shown in Figure 2.20. Most SBR contains 20–23% styrene and is used in the manufacture of tires for motor vehicles, but it has a serious limitation because of its absorption of many solvents and subsequent swelling. NBR contains 18–45% acrylonitrile and was developed to produce a solvent-resistant elastomer. Its main applications are for fuel hose and oil-resistant gaskets.

Polychloroprene or neoprene rubbers are similar to polyisoprene with the exception that a methyl group is replaced by the chlorine atom, as shown in Figure 2.21. Ordinary SBR and NBR and natural rubbers deteriorate by oxidation, which embrittles them. The presence of chlorine in the neoprene rubbers provides oxidation resistance, useful in applications such as pulley belts, automotive seals, and diaphragms. All of the synthetic rubbers are strengthened by fillers, the most common of which is carbon black. Such fillers are not only beneficial to the properties, but lower the cost of the materials as well.

**Silicones** are a special class of polymers based on the repetition of silicon (which has very similar characteristics to carbon) and oxygen in the polymer chain,

**Figure 2.18**
Cross-linking in vulcanized rubber where cross-links, typically sulfur atoms, form between cis-polyisoprene long-chain molecules (After W. G. Moffatt, G. W. Pearsall, and J. Wulff, *The Structure and Properties of Materials,* Vol. 1: *Structure,* John Wiley & Sons, 1964.)

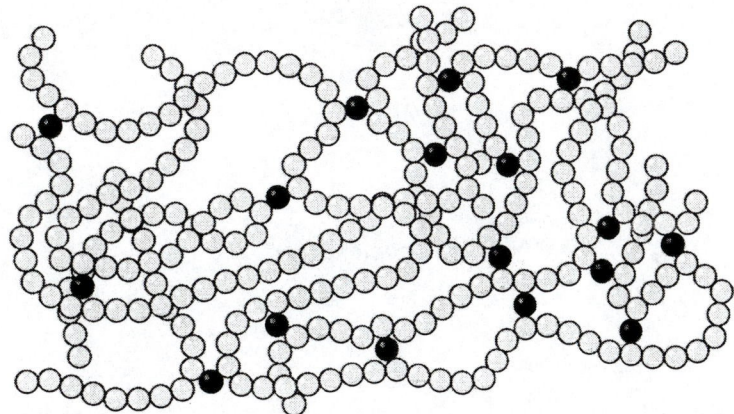

**Figure 2.19**
Stress-strain behavior of vulcanized and unvulcanized natural rubber
(After M. Eisenstadt, *Introduction to Mechanical Properties of Materials*, Macmillan, 1971.)

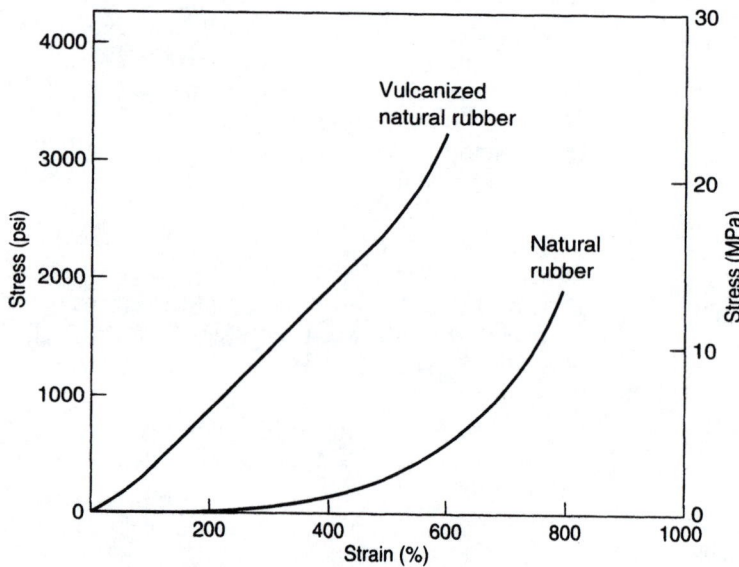

**Figure 2.20**
Mers of SBR and NBR synthetic rubbers

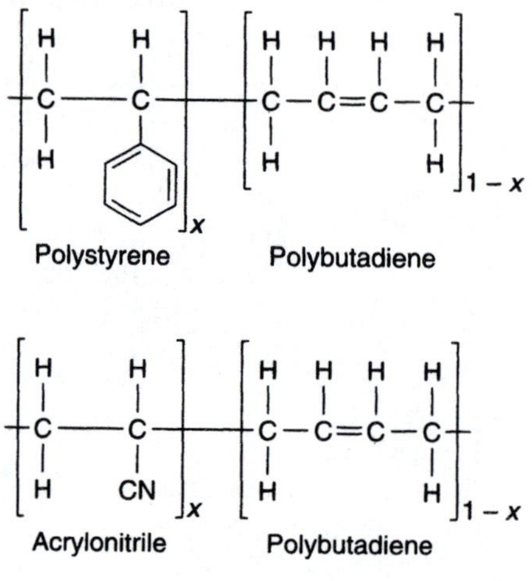

Polystyrene    Polybutadiene

Acrylonitrile    Polybutadiene

**Figure 2.21**
The polychloroprene mer

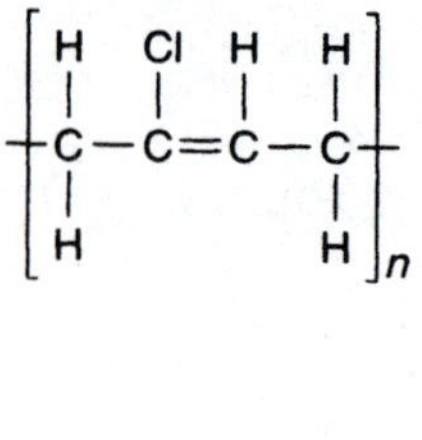

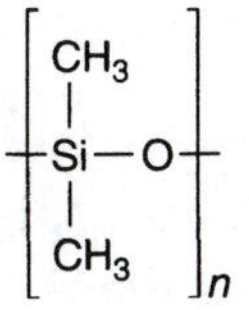

**Figure 2.22**
The polydimethylsiloxane mer

$$\left[\begin{array}{c} CH_3 \\ | \\ -Si-O- \\ | \\ CH_3 \end{array}\right]_n$$

called a *siloxane* chain. The most common elastomer is polydimethylsiloxane; this silicone monomer is shown in Figure 2.22. The siloxane chain is cross-linked by a number of different organic and inorganic chemicals.

A great deal of energy is needed to produce silicones, thus limiting their application because of high cost. Nevertheless, silicones are available as fluids, compounds, resins, lubricants, and elastomers because of their high-temperature capabilities and superior electrical and sealing characteristics. They are used extensively in wire and cable insulation, as gaskets, and in aerospace applications. Silly Putty, a novelty toy, is a silicone. Fluid silicones are used as mold-release agents and in lens-cleaning tissues. Silicone compounds such as RTV (room-temperature vulcanizates) are for vacuum sealants and for potting of electronic components, resins for high-temperature paints, lubricants for high-temperature applications, and elastomers for adhesives and sealants, usable over a wide temperature range. Table 2.3 summarizes some properties of common elastomers.

## Case Study 2.4

### Designing with Silicone Sealant

The Industrial Revolution in America started with the textile industry in Lowell, Massachusetts, which today is the site of a national historic park. Many of the old mills have been renovated for use as conference centers and museums. The architectural plans for the Manufacturing Museum called for a new roof that was to be sealed with silicone. Specifications required three layers—a primer, base coat, and top coat—with a total dry film thickness of 0.022 in. The finish was required to be applied over flashing at the base of plexi-

**Table 2.3**
Typical properties of selected elastomers

| Elastomer | Density (g/cc) | UTS (ksi) | Maximum temperature (°F) |
|---|---|---|---|
| Natural rubber | .93 | 3.0 | 180 |
| SBR | .94 | 3.0 | 180 |
| NBR | 1.00 | 0.8 | 250 |
| Neoprene | 1.25 | 3.5 | 240 |
| Silicone | 1.35 | 1.0 | 315 |

glass domes used for indirect lighting of the museum display areas. This finish was textured by dispersing aluminum oxide or silica aggregate over the silicone before final curing was complete.

The roof subsequently leaked, damaging the internal remodeling. Leakage was particularly bad around the domes, which caused extensive delays in installing the antique textile machinery and the inauguration of the museum. Samples of the roofing were brought to the university for failure analysis. Simple examination in a stereozoom microscope showed three types of holes through the silicone thickness—tears, inadequate thickness, and penetration by the aggregate. Figure 2.23 illustrates a tear through the silicone where the thickness was as thin as 0.011 in., half the specified thickness. Figure 2.24 shows inadequate silicone application over a fabric; the weave of the fabric is visible. Cross sections were mounted and polished, then measured optically in a metallurgical microscope. It was found that the specified thickness of silicone was applied in most cases, but this was compressed by the aggregate particles, particularly the largest ones. Unfortunately, the samples were not correlated with the locations from which they were removed, but costly repairs were undertaken and the responsibility was shared by the general contractor and the architectural firm.

**Figure 2.23**
Tear in a section of silicone
roofing (10×)

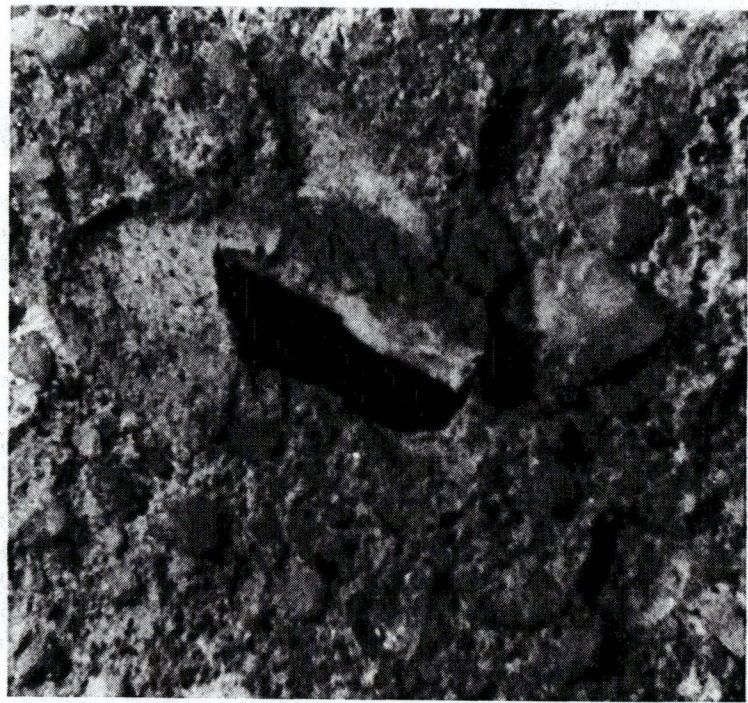

**Figure 2.24**
Area of fabric with thin
silicone layer (10×)

# 2.5.4 Miscellaneous Polymers Better Classified by Shape or Utilization

There are many polymers that do not easily fit into the classifications of thermoplastic, thermosetting, or elastomeric because they were developed for specific qualities or for specific uses. Yet they deserve to be recognized, and we will do so here.

Many of these materials are used as coatings for decoration and protection of wood surfaces. The **paints** that are typically applied to a surface by brushing, rolling, or spraying consist of pigment for color, a solvent such as water for latex paints or linseed oil for oil-base paints, and a polymer. The solvent extends the polymer, making it easier to apply the paint. When the solvent evaporates, a polymerized film entrapping the pigments is left. Important polymers used for paints are either thermoplastic or thermosetting and include vinyls, epoxies, rubbers, acrylics, silicones, and phenolics, all of which have already been discussed. There are other important polymers, such as alkyds and urethanes, that have not been mentioned. **Alkyds** are unsaturated polyesters that are modified with acids, usually the fatty acids of linseed, soybean, or other oils. Naturally, they are used in oil-base paints, but they can also be used as molding compounds that polymerize under heat and pressure for use in electrical applications where the cheaper phenolics might not be suitable. **Polyurethane** coatings are polymerized from the reaction of polyisocyanates (–NCO–) and polyhydroxyl (–OH–) groups; the repeating unit in polyurethane is

–NHCOO– or –NHCO–. These coatings are best known for their toughness, abrasion resistance, and flexibility, which makes them attractive for outdoor as well as indoor use.

Polyurethane is also produced as rigid fibers, which are cut up into the bristles used for paintbrushes. But resourceful chemists at DuPont developed a **block copolymer** in which the rigid polyurethane block alternated with a soft polyether block. We see these fibers frequently in sportswear, but know the material better as **spandex**. Spandex fibers have an elongation of more than 500%.

## Summary

Polymers or plastics are organic materials that have exploded onto the marketplace in the second half of the twentieth century. They are high molecular weight, long-chain molecules that are classified as linear, or aliphatic, and as aromatic when rigid rings appear in the long chain. A solid is built up by polymerization of the individual units, or mers, by addition or condensation processes. Although plastics are inorganic long-chain molecules, they can have some order, which we refer to as crystalline; of course, the properties of the plastics are related to the degree of crystallinity. A convenient method of classifying the many polymers is by the terms thermoplastic and thermosetting. Thermoplastic polymers can be reshaped, whereas thermosetting polymers cannot. Typical thermoplastic materials are polyethylene, polystyrene, and others with the vinyl linkage, such as acrylics and polycarbonates. Typical thermosetting polymers include phenolics, polyesters, and epoxies. Elastomers are a special category of polymers that can elongate more than 200% and fully recover when unloaded. They include natural rubber, synthetic rubbers, such as styrene butadiene and nitrile rubbers, and silicones. There are other polymers that we have classified by their use, such as alkyd used in oil-base paints and polyurethane used for coatings and as rigid paintbrush bristles.

## Terms to Remember

| | |
|---|---|
| ABS (acrylonitrile-butadiene-styrene) | condensation |
| acrylic | copolymer |
| aliphatic | elastomer |
| alkyd | epoxy |
| aromatic | fillers |
| atactic | films |
| Bakelite | graft copolymer |
| block copolymer | gutta-percha |
| coatings | isotactic |

Kevlar

melamine formaldehyde

mer

monomer

NBR (nitrile rubber)

neoprene

nylon

Orlon

paint

PAN (polyacrylonitrile)

PET (polyethylene terephthalate)

phenolic

plastic

plasticizer

polyacetal

polyamide

polycarbonate

polyester

polyethylene

polymer

polymethyl methacrylate

polyolefin

polypropylene

polystyrene

polyvinyl chloride (PVC)

polyurethane

resin

rubber

SAN (styrene acrylonitrile)

SBR (styrene butadiene rubber)

silicone

spandex

syndiotactic

terpolymer

thermoplastic

thermosetting

vulcanize

# Problems

1. Explain in your own words what polymers are and how you would classify the many polymers available in the marketplace.
2. Explain in your own words what polymerization is.
3. Select and describe the composition of a typical polymer you would find in the form of a film, a fiber, a coating, and in bulk form.
4. Explain in your own words the role of additives in polymers. Give an example of each.
5. Explain in your own words what crystallinity means when referring to polymer materials. How does crystallinity affect properties?
6. Explain how you would prevent the failure of the clamp discussed in Case Study 2.1.
7. Describe some products made of polyamides, polycarbonates, and phenolics that you have used.
8. Describe some elastomeric products that you have used.
9. Explain, with examples, how polymers are used in electrical and electronic equipment.
10. Explain, with examples, how polymers are used in the construction industry.

# 3

# Polymer Processing

In general, the properties of polymers, ceramics, and composites are not influenced by **processing**. Properties of the parts produced will be controlled by the polymer itself or, in the case of thermosets, by the cross-linking that takes place during the processing. In this chapter, we will examine the many processes used to shape polymer materials, processes that form the multitude of shapes and designs of products. Some of them are similar to those used in fabricating metal parts but do not require high temperatures, and others are very different.

A convenient way to address the processes used for plastics is to separate those used for thermoplastic polymers from those used for thermosetting polymers. Thermoplastic materials are usually heated to above their melting or softening temperature and shaped (or reshaped), whereas thermosetting materials undergo final polymerization by cross-linking during processing. Although we will adopt this approach, keep in mind that it is only a convenience. We will also examine methods to expand and shape foams, laminates, and films that might be either thermoplastic or thermosetting.

# 3.1 Processing of Thermoplastic Polymers

The three major processing methods used for shaping **thermoplastic** materials are extrusion, injection molding, and thermoforming. In all of these, the most important material properties that we must consider are the **viscosity** (resistance to flow) and flow characteristics. For polyethylene, we use the **melt index (MI)** as a guide, but for all other polymers, we use **melt flow rate (MFR)**. These parameters are affected by molecular weight; high MI or MFR results for low molecular weight, but properties are reduced. Table 3.1 reflects the changes in properties of polyethylene related to changes in molecular weight and degree of **crystallinity**. The MI or MFR is a measure of the amount of material that is extruded in a specific time through a specific small orifice at a pressure of 43.5 psi at a specific temperature above the melting or softening point of the polymer.

A high value of the melt index indicates a low viscosity material. If we use a polymer with a high melt index or melt flow rate, less molding pressure is needed, and warpage and shrinkage are decreased, but toughness and flexural characteristics are reduced and there is less resistance to stress cracking while shaping the polymer. Of course we seek many other tests that can help us improve productivity and quality assurance, but such techniques (e.g., shear capillary rheometers) relate more to mold design than to the processing.

**Table 3.1**
Effect of crystallinity and melt index on properties of polyethylene

| Property | Change if crystallinity increases | Change if melt index increases |
|---|---|---|
| Melt viscosity | Higher | Lower |
| $T_g$ | Much higher | Lower |
| Surface hardness | Higher | Slightly lower |
| Resistance to mold sticking | Higher | Slightly lower |
| Yield strength | Much higher | Slightly lower |
| Elongation | Lower | Lower |
| Flexural stiffness | Much higher | Slightly lower |
| Resistance to brittleness at low temperature | Lower | Lower |
| Resistance to environmental stress cracking | Lower | Lower |
| Shrinking | Higher | Lower |
| Warping | Slightly higher | Lower |

# 3.1.1 Extrusion

**Extrusion** for polymers is the same process as that used in metal fabrication. However, for polymers the process can be continuous because granulated polymer material can be fed without interruption from a hopper into an auger or screw that carries it through a heating chamber. The action of the screw compacts the polymer as it is heated to temperatures at which the polymer is molten. This plasticated material (which has been melted and mixed), or extrudate, is then forced through a die, as shown in Figure 3.1. In this figure, the extruder is set up to apply electrical insulation to a copper wire. In many extrusion processes, the hot polymer can be formed further after leaving the die by stretching, postforming, or spinning before it cools. Typical extruded products are pipe (the molten polymer is forced between an outer die and a mandrel), shaped profiles (e.g., gutters), sheet and film (including blown film), fibers, and coatings such as the insulation for electrical wiring.

There are many factors that control the quality of extruded products. In the extruder, the most important component is the screw itself, which is divided into the feeder section, the transition section where melting takes place, and the metering section where the product is pushed at a uniform rate and pressure through the die. Of course, the die configurations and therefore the screw configurations vary greatly. For example, we can extrude round pipe using a mandrel and also extrude wide sheet product that requires a T-shaped or coat hanger-shaped die whereby extrudate is fed to the center of the die and formed outward to fill the width before being extruded from the die.

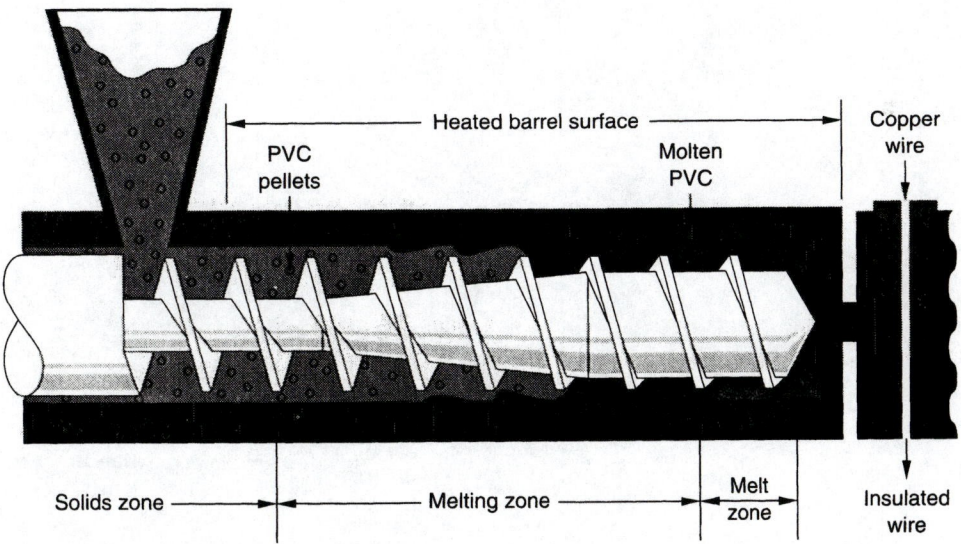

**Figure 3.1**
A plasticating extruder for insulating copper wire

A special version of the extrusion process is blown-film extrusion, which is used to produce low-cost films and thin-film containers. In blown-film extrusion, extrudate is forced around a mandrel and through a die. The thin, tubular film is enlarged by blowing air through the mandrel orifice to expand the molten film. When the **blown film** has expanded to the desired wall thickness, it is cooled by air from a water-cooled ring around the die. Blown-film extrusion is shown schematically in Figure 3.2.

Polymer fibers are also made by the extrusion process. This can be accomplished traditionally by plasticating and extruding the polymer through tiny holes of a special die called a **screw spinneret** that can rotate to form multifilamentary thread that is wound directly onto spools. Nylon threads are produced in this way. Acrylics such as Orlon are produced by solvent spinning, whereby the cold polymer is dissolved and extruded through the spinneret, then solvents are evaporated before spinning is completed. Wet spinning, used for some materials such as rayon, begins with a dissolved polymer that is extruded through the spinneret into a coagulating bath where it is polymerized into a solid filament, then spun. These methods for producing spun plastic fibers are shown in Figure 3.3.

**Figure 3.2**
Schematic process for blown-film extrusion

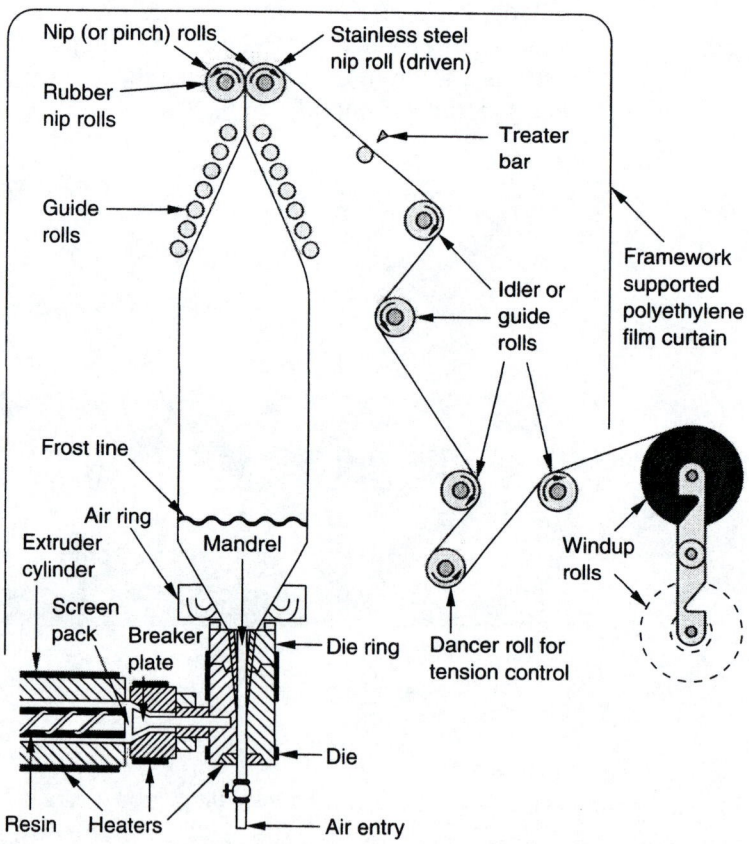

# 3.1.2 Injection Molding

In **injection molding**, granular polymers are heated and forced, or injected, into a mold cavity, cooled, and removed. The process is similar to metal die casting in most respects and is used to produce many items, from toothbrush handles to automobile bumpers. Mold design requires feeding gates and runners that are removed, reground, and recycled, so the process is extremely efficient. All thermoplastic polymers (except those with high viscosity like fluoroplastics, polyimides, and some aromatic polyesters) can be injection molded and many thermosetting polymers can be injection molded as well. Advantages of injection molding are high production rates, ability to inject around inserts, little or no finishing requirements, and recycling capability. The only major disadvantage is the high cost of equipment and of molds.

Modern injection-molding machines use the reciprocating screw for plastication, the same as for extrusion, but the screw is stopped and then used as a plunger. Plunging injects the hot polymer into the mold where it takes the shape of the cavity, as shown schematically in Figure 3.4. When cooled, the parts are removed and gates,

**Figure 3.3**
Extrusion processes for spun polymer fibers: (a) melt spinning, (b) solvent spinning, (c) wet spinning (Reproduced by permission. *Industrial Plastics: Theory and Application,* 2nd ed., by T. L. Richardson. Delmar Publishers, Inc., Albany, NY. Copyright 1989.)

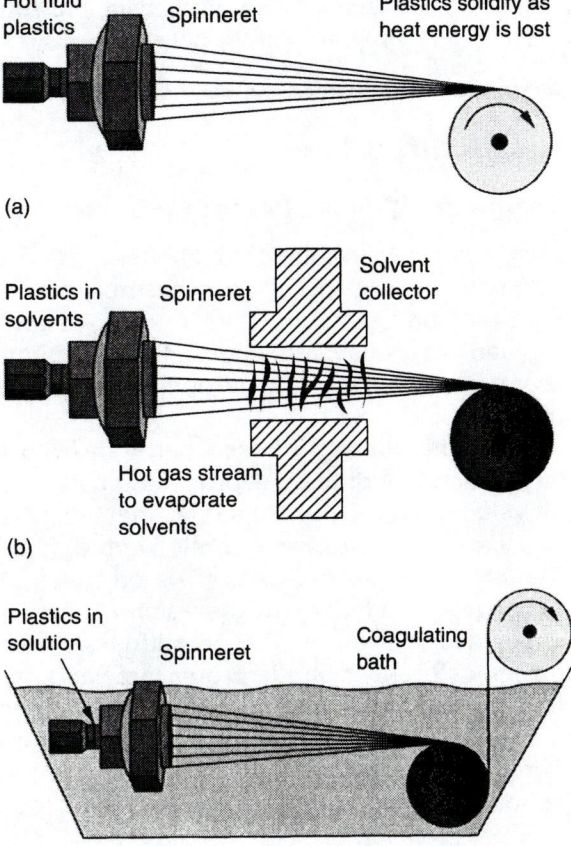

Hot fluid plastics — Spinneret — Plastics solidify as heat energy is lost

(a)

Plastics in solvents — Spinneret — Solvent collector

Hot gas stream to evaporate solvents

(b)

Plastics in solution — Spinneret — Coagulating bath

(c)

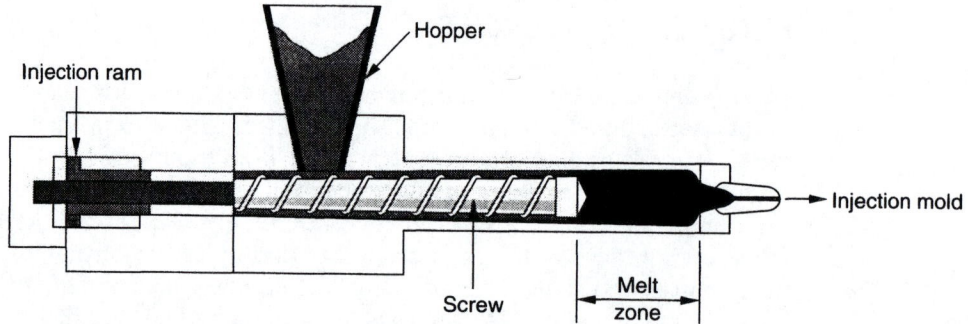

**Figure 3.4**
Schematic diagram of reciprocating screw injection-molding machine

runners, and excess material are removed and reground for recycling. Parameters that must be controlled are pressure, mold and material temperature, flow and shear properties of the polymer, and melt and flow control, among others.

## Case Study 3.1

### Change the Material, Change the Vendor, but Change Both?

UML Industries has successfully been using polypropylene connectors that were injection molded by Inject, Inc., for an assembly that joined the polypropylene to a steel framework with nuts and bolts inserted to 10 in.-lb torque. MoldPro Industries offered an injection-molded polyphenylene oxide (PPO) substitute at a comparable price, even though PPO is more expensive than polypropylene. UML decided to consider changing to the PPO connectors.

Examination of unused parts showed that the wall thickness of the PPO part was larger than that of the polypropylene part by as much as 200% around bolt holes, but overall volume was nearly the same. Trial lots showed a 20% rejection rate of the new material because of cracking when bolts were tightened, but this was reduced to about 5% when the torque for insertion was lowered to 4 in.-lb. Careful inspection in a stereozoom microscope revealed that many cracks and checks were present in incoming PPO material. It was concluded that the increased stiffness of the PPO polymer combined with feeding gate design differences and improper pressure and temperature controls were the major cause for rejection and that the use of polypropylene should not yet be abandoned.

Although economic savings is an admirable goal, it cannot be a substitute for quality. Careful inspection and testing in this case prevented a change in vendors that could have seriously damaged the reputation of UML Industries.

# 3.1.3 Thermoforming

**Thermoforming** is equivalent to a secondary process in metalworking. It begins with a sheet or film made by other processes such as extrusion, then reforms it in a mold under heat and pressure to a final shape. Such diverse products as auto bodies, windshields, advertising signs, and food containers are made in this way. Even the shrink-wrapping of commercial and consumer products is an example of thermoforming.

There are many variations of thermoforming. The simplest and most versatile is **vacuum forming,** whereby the polymer sheet is heated and the cavity is evacuated. Atmospheric pressure is sufficient to force the sheet into the cavity. **Drape molding** is similar to this method except that the hot plastic is stretched mechanically over the mold. Figure 3.5 illustrates this process.

In many thermoforming presses, sheet is transferred from a payoff roll through a station where thermoforming takes place between matched heated molds, then through a trimming operation where parts are removed; remaining scrap is coiled

**Figure 3.5**
The drape-molding process: (a) clamped heated plastics may be pulled over the mold, or the mold may be forced into the sheet; (b) once the sheet has formed a seal around the mold, a vacuum is drawn to pull the plastic sheet tightly against the mold surface; (c) final wall thickness distribution in the molded part

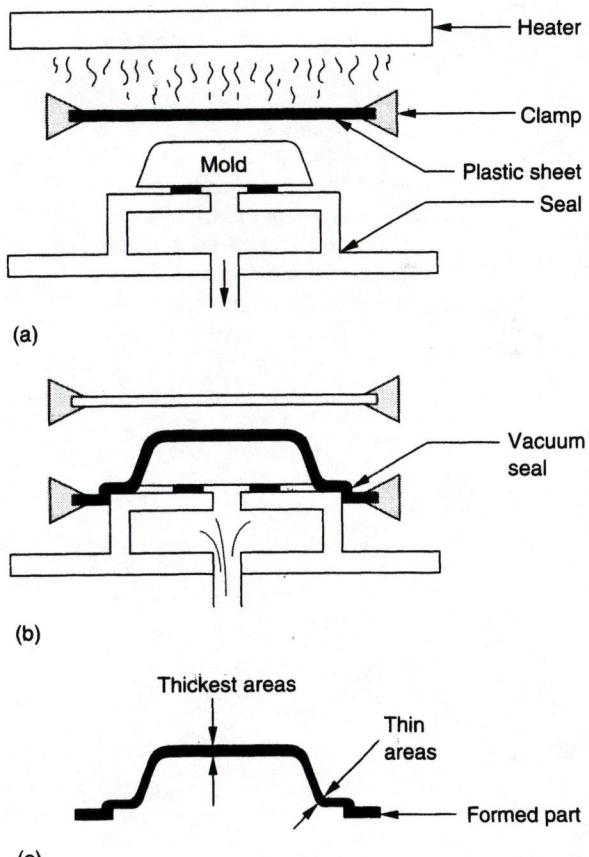

onto a take-up roll that can be recycled. Matched mold forming is shown schematically in Figure 3.6. Large deformations, or deep draws, are used to form luggage, auto parts, and housings for electronic and other equipment. Vacuum snap-back thermoforming is used for such complex shapes. In this process, polymer sheet is heated and sealed over the top of a vacuum box that is subsequently evacuated, drawing the sheet into it. A male plug that has vacuum drawn through it is lowered and the original vacuum is released, and air is reintroduced, snapping the sheet against the male die. Other methods of thermoforming include straight mechanical forming of heated plastics over mechanical forms, free forming that uses air pressure to expand a clamped sheet, and air slip forming, which combines the free forming with snap-back features of vacuum snap-back forming.

**Figure 3.6**
Matched mold forming: (a) the heated plastic sheet may be clamped over the female die, as shown, or draped over the mold form; (b) vents allow trapped air to escape as the mold closes and forms the part; (c) distribution of materials in the product depends on the shapes of the two dies; (d) male mold forms must be spaced at a distance equal to or greater than their height or webbing may occur

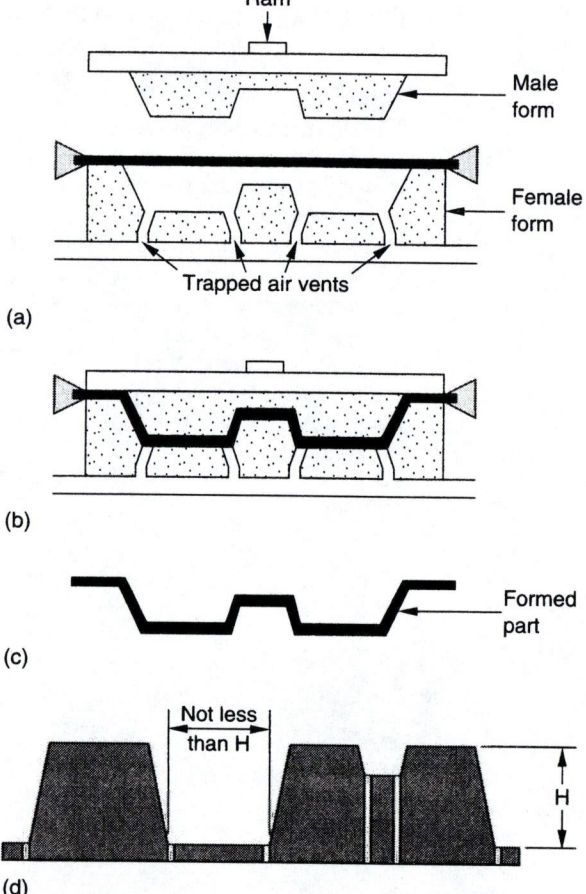

### Case Study 3.2

#### Polycarbonate Processing

Polycarbonates are best known for bullet-resistant or shatter-resistant glass applications, consequences of their outstanding impact resistance and durability. Polycarbonates are extremely versatile, however, and are used for mechanical gears, automobile headlamps, bumpers, instrument panels, housings for electronic equipment such as printers and copiers, and for sports equipment such as football helmets. A high purity grade is even used for manufacturing compact discs and optical memory discs.

Polycarb, Inc., has been in the forefront of polycarbonate manufacture, developing extrusion, injection-molding, and thermoforming processes specifically designed for polycarbonates. They found moisture control to be critical and predrying necessary because residual moisture reacts at the processing temperature, reducing the molecular weight and the toughness. Sheet had to be produced by extrusion at temperatures of 450–650°F. Polycarb also found the temperature of cooling rolls to be important; if they were too cold, warpage resulted due to differential contraction and if they were too hot, the sheet sagged. Actual roll temperature, of course, depended on the thickness of the sheet being extruded. They found also that the rolls determine the surface finish, which is preserved by application of a paper backing during the processing.

Most housings are injection molded by Polycarb because shrinkage problems are reduced by adding glass fillers. For example, addition of 40% glass to polycarbonate reduces the thermal expansion coefficient from $3.75 \times 10^{-5}$ in./in./°F to $1 \times 10^{-5}$ in./in./°F. Polycarb, Inc., also cold-forms polycarbonates mechanically and can, of course, thermoform them. For example, they use free forming to produce skylights and transparent domes.

# 3.1.4 Miscellaneous Processes for Thermoplastic Polymers

There are many other methods to process thermoplastic materials besides the major ones we have just examined. For example, film casting resembles the Fourdrinier processing of sheet paper (see Chapter 10). A polymer is cast on a smooth conveyor and solvent is evaporated by heat, leaving only the **cast film** remaining. Polycarbonates, polyvinyl chloride, and some cellulosic polymers can be film cast, but perhaps the most common example is the nonstick Teflon (polytetrafluoroethylene) surfaces of frying pans.

The miscellaneous processes that account for the largest volume of thermoplastic products, however, are blow molding, calendering, laminating, and expansion processes. **Blow molding** is used to shape thermoplastic materials that have high hot strength combined with good stretch characteristics. Blow-molded products such as barrels, jars, milk bottles, and many others are commonly made from low-density polyethylene, polypropylene, polyvinyl chloride, polycarbonates, and polystyrene. **Calendering** consists of hot blending all components in a continuous kneader called a Banbury mixer, passing the blend through hot rollers to thin the product to the desired thickness, then passing it through a final series of cold rolls. This process, shown schematically in Figure 3.7, is most frequently used for polyvinyl chloride sheet. **Laminating** is used for BOTH thermosetting and thermoplastic polymers, but in this section we will only discuss thermoplastic lamination, such as the three-ply laminate of polyethylene-polypropylene-polyethylene used for packaging bread. Expansion processes reduce the density of the polymer for applications such as weight reduction, cushioning, flotation, and insulation, among others. There are a number of methods to expand polymers; some of them use chemical blowing agents, dissolution of a gas that later evaporates and expands the polymer, mechanical whipping, and chemical reaction. Let's look at these processes in more detail.

*Blow molding* was developed for producing glass bottles long before polymers were known. A gob of molten glass was pressed into a mold, removed, then air was blown into the *parison*, forcing it against a mold. The method remains quite similar for stretch blow molding of thermoplastic polymers, such as the 2-liter polyethylene terephthalate (PET) bottle shown in Figure 3.8. Injection molding is used to make a *preform* (also called the parison), which is transferred, reheated, and expanded by blowing air into it. Blow molding can also incorporate extrusion, whereby the parison is extruded around a vented mandrel, then air is injected into the mandrel. Extrusion blow molding has some advantages because of reduced die costs, reduced handling, and elimination of transfer time.

**Figure 3.7**
The calendering process
(Reproduced by permission.
*Industrial Plastics: Theory
and Application,* 2nd ed.,
by T. L. Richardson. Delmar
Publishers, Inc., Albany, NY.
Copyright 1989.)

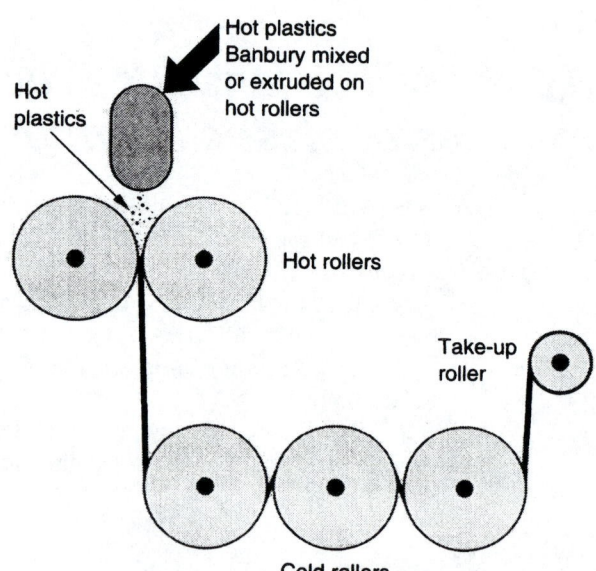

*Calendering* is usually associated with the textile industry. Polymer sheet such as polyvinyl chloride, for example, is embossed to give the appearance of leather or other desired textures. Calendering is also used to blend different color polymers to produce multicolored floor tiles with smooth walking surfaces and rougher bottom surfaces for adhesion to the underlying floor. A unique product is fibrous polyolefin, used extensively as housewrap or in mailer envelopes. Fibers are extruded, cut, and then mechanically defibrillated, similar to defibrillation of wood fibers in papermaking (see Chapter 10). They are then calendered, flattening the fibers and providing adhesion to produce the tough, open structure required for such applications. Figure 3.9 illustrates the open structure for a Tyvek mailing envelope, one of many grades of this product developed by E. I. DuPont.

**Figure 3.8**
Stretch blow molding:
(a) injection molding of
parison, (b) parison
ejection, (c) stretch blow
molding of reheated
parison

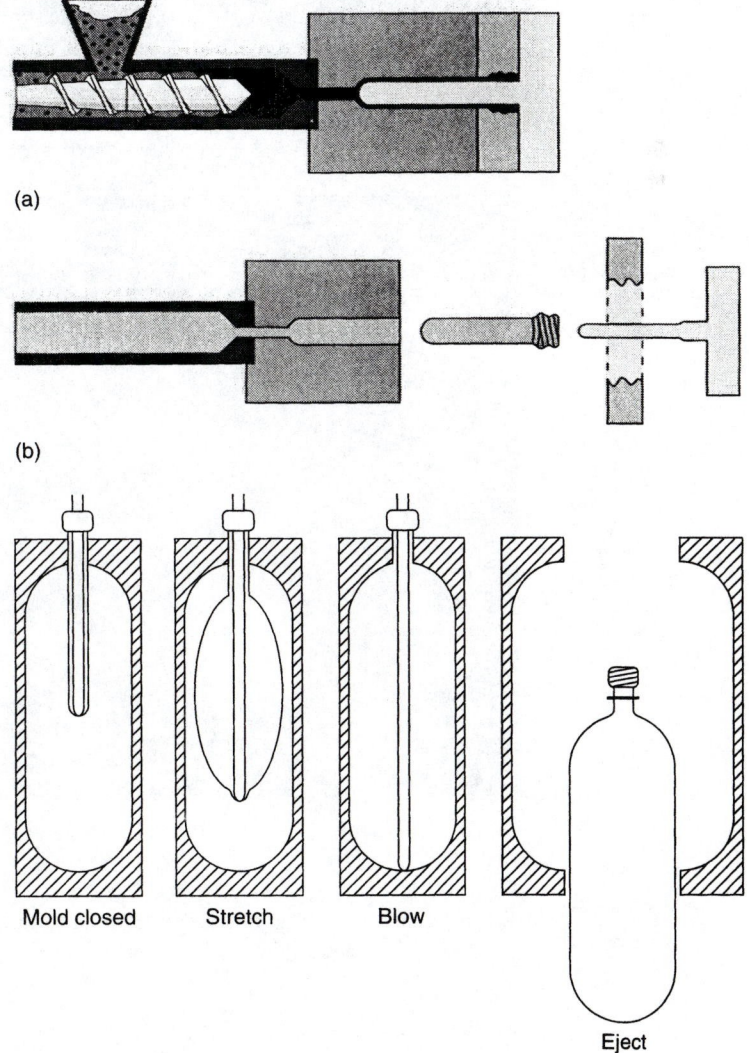

(a)

(b)

Mold closed     Stretch     Blow

Eject

(c)

(a)

(b)

**Figure 3.9**
Tyvek mailing envelope produced by calendaring polyolefin fibers:
(a) 250×, (b) 2500× (Courtesy of Drew Killins.)

## Case Study 3.3

### It's Fashionable to Recycle

Technology has reached the stage where the plastic 2-liter bottles we leave at the curbside can be reincarnated as our latest fleece sportswear or the insulation in new sleeping bags. The rough recycled fibers once usable only for carpeting can now be made into finer fibers that are much softer. The fineness of fibers is measured by a unit called the **denier**, which is the mass in grams of a fiber 9000 meters long. (Clothing fibers range from 2 to 10 denier, carpet fibers from 15 to 30 denier.)

How this technology developed is a story of the environment and economy. We produce about 2½ million PET bottles every hour in the United States alone. When discarded, they occupy 30% of our landfill capacity, a problem that is being solved now by recycling. PET bottle parts are separated from non-PET parts and inserted into granulators that produce flakes. The flakes must be very clean and free of all contaminants, therefore extensive cleaning is required. The cleaned flakes are then heated, dyes are added, and the molten material is filtered into strips that resemble spaghetti. The melt filtration eliminates minute contaminants that might remain. Strips are reheated, stretched, and cut into 2-in. lengths, called *staple fibers*. After these staple fibers are carded, they can be drawn into yarn for weaving or can be collected and felted using a needle punch for nonwovens. By being able to control the denier, current technology has made it possible to convert blow-molded PET into the low-denier polyester fibers needed for the fleece outerwear so popular nowadays. Figure 3.10 shows the fine denier of the polyester fibers required for this product.

*Laminations* are layered sheet composite materials in which the layers are bonded together by high pressure and high temperature. One of the best known laminates is Formica, which was originally developed as an electrical insulation material, an application surpassed many times over by its use in construction and furniture today. For thermoplastic laminations, multiple extruders are combined to feed extruded sheet and adhesives, continuously forming a multilayered composite sheet. Individual layers may be colored or composed of different materials and can be joined to foil, paper, or fabric.

*Foamed plastics* are not only versatile with respect to their many applications, but also in their processing. They can be foamed in place, injection molded, extruded, thermoformed, laminated, and cast to shape. They can be flexible, rigid, or anywhere in between, and can have open cells to be absorptive or closed cells for reduced density. Expanded thermoplastic polymers include polystyrene, polyvinyl chloride, and polyethylene.

**Figure 3.10**
Nonwoven polyester fleece (250×)

Expanded polystyrene foam, known well by the Dow Chemical trade name of Styrofoam, is extruded from polystyrene pellets combined with a blowing agent. Extruded **expanded foam** can be thermoformed into a variety of containers, trays, or cartons. **Expandable styrene** parts (e.g., a coffee cup) are usually made by a two-stage process, whereby polystyrene beads are pre-expanded (usually by steam heat), equilibrated, then transferred to the shaped mold where additional steam expands them further to fill the mold and sufficiently bond the expanded beads for use. These expandable polystyrenes can be distinguished from expanded polystyrene foam because the round bead lines of the expanded foam are distinct from the thin inner walls, as shown in Figure 3.11.

Polyethylene and polyurethane can be expanded in this same fashion, with a surface skin formed in air after passing through a die. Figure 3.12 illustrates the surface and internal wall structure of an expanded rigid polyethylene foam as well as a soft polyurethane foam made in this manner. Foamed materials are processed as sheet, which is used for backing of carpeting and other materials, or the foam can be applied directly to a surface by using mechanically frothed foam or by later laminating. Many times, the vinyl is expanded after a skin is formed, producing a soft foam material with skin on both sides, as shown in Figure 3.13.

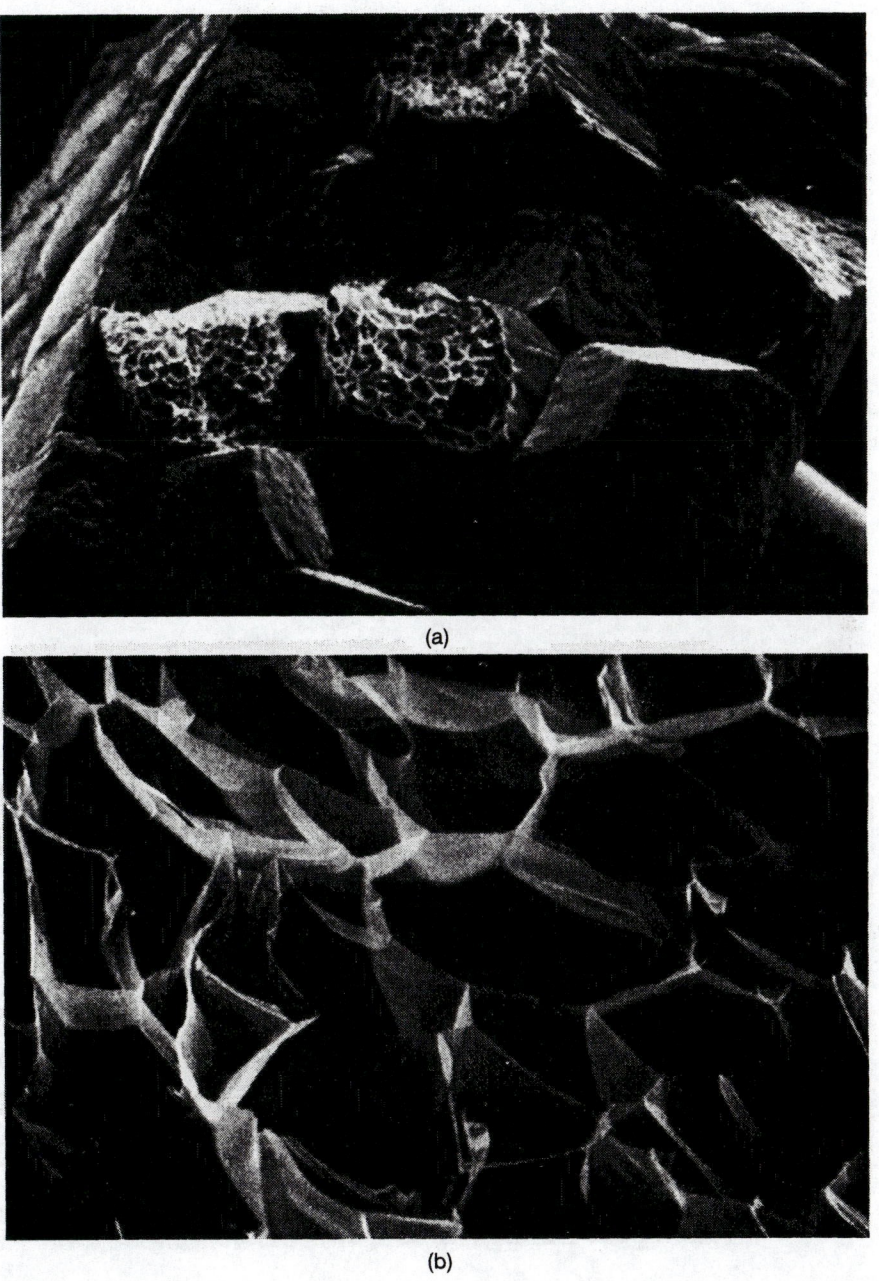

(a)

(b)

**Figure 3.11**
Expanded polystyrene: (a) boundaries of pre-expanded beads (25×), (b) inner
walls of re-expanded beads (250×)

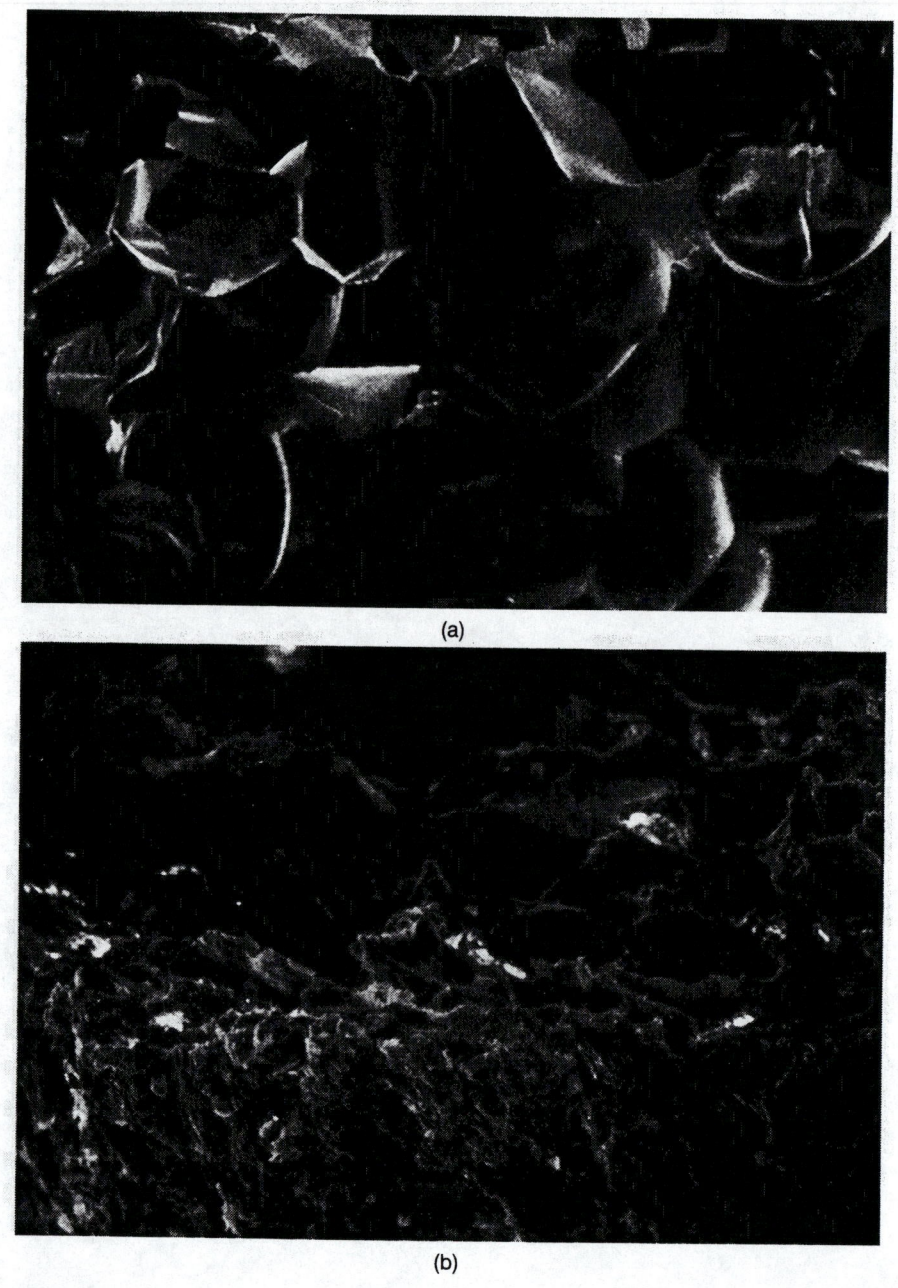

(a)

(b)

**Figure 3.12**
Blown foams with skin formed in air: (a) polyethylene (11×), (b) polyurethane (125×)

**Figure 3.13**
Method for forming flexible vinyl foam with skin on both sides
(Reproduced by permission. *Industrial Plastics: Theory and Application,* 2nd ed., by T. L. Richardson. Delmar Publishers, Inc., Albany, NY. Copyright 1989.)

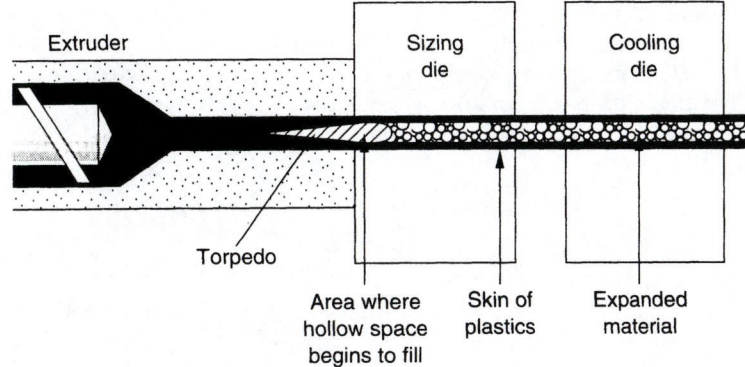

# 3.2 Processing of Thermosetting Polymers

Many processes can be used for **thermosetting** materials as well as thermoplastic materials, but some are better known for major use with one or the other type of material. We usually associate extrusion, thermoforming, and injection molding with thermoplastic materials, for example, and compression molding, transfer molding, pultrusion, and casting with thermosetting materials. Lamination and expansion processes are frequently used for either classification.

## 3.2.1 Compression Molding

**Compression molding** is not unique to thermosetting polymer processing—it is one of the oldest methods used to shape all kinds of materials, from ceramics (molded bricks made the brickmakers of Babylon famous in ancient times) and metals (the process is similar to closed-die forging) to modern composite materials. In thermoforming materials, compression molding is used almost exclusively for production of polytetrafluoroethylene or for specialty ultra-high molecular weight, ultrafine polyethylene (see Chapter 6), but rarely for others. Thermosetting materials that are compression molded include phenolics, melamines, epoxies, and composites.

In compression molding, the thermosetting material, which might be loose or premolded, is placed into the cavity of a die, then heat and pressure are applied, curing the polymer. Figure 3.14 illustrates the open mold with a preform in place and the closed mold with flashing formed by excess material. Pins in the lower portion of the mold are for ejection of the completed part. Laminates are also prepared by compression molding, but no die is used, quite like open-die forging processes for metal. For example, electronic printed wiring boards are made up of layers called *prepreg* (woven fiberglass that has been coated and partially polymerized). These are stacked, some with printed copper wires, then polymerized by heat and pressure in

**Figure 3.14**
Compression molding:
(a) open mold with preform
in place, (b) closed mold
showing formation of flashing

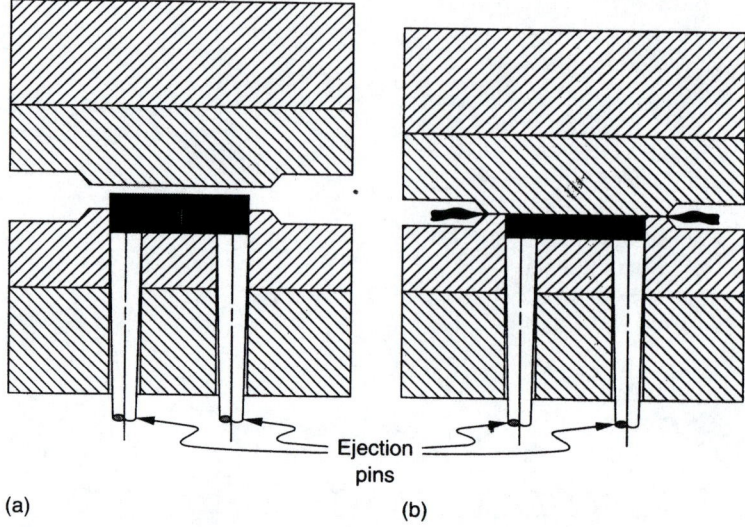

Ejection
pins

(a)                                (b)

an open press. Some of the advantages of compression molding include low tooling costs and little waste material, but only relatively simple parts can be molded and scrap cannot be recycled.

## 3.2.2   Transfer Molding

In compression molding, the preformed or loose plastic resin is placed directly into the mold cavity. This is not true for **transfer molding** because the resin is placed into a chamber, then transferred into a number of mold cavities through a system of gates and runners much like the feeding system for injection molding thermoplastics. In comparison to compression molding, more complex parts can be made, there is less flash, and there are shorter molding times, but there is more waste and mold costs are higher. Figure 3.15 shows the sequence in a simple transfer molding operation. Transfer molding is used in the microelectronic industry to encapsulate integrated circuits in plastic packages that are later joined to other components via soldering to the printed wiring of a laminated printed wiring board.

## 3.2.3   Pultrusion

**Pultrusion** is unique to thermosetting resin processing when reinforcements are incorporated into the resin matrix; it is similar to metal wire drawing but for the fact that it is a continuous process and curing takes place while being drawn through the die. Pultrusion dies, therefore, are much longer and are heated in comparison to

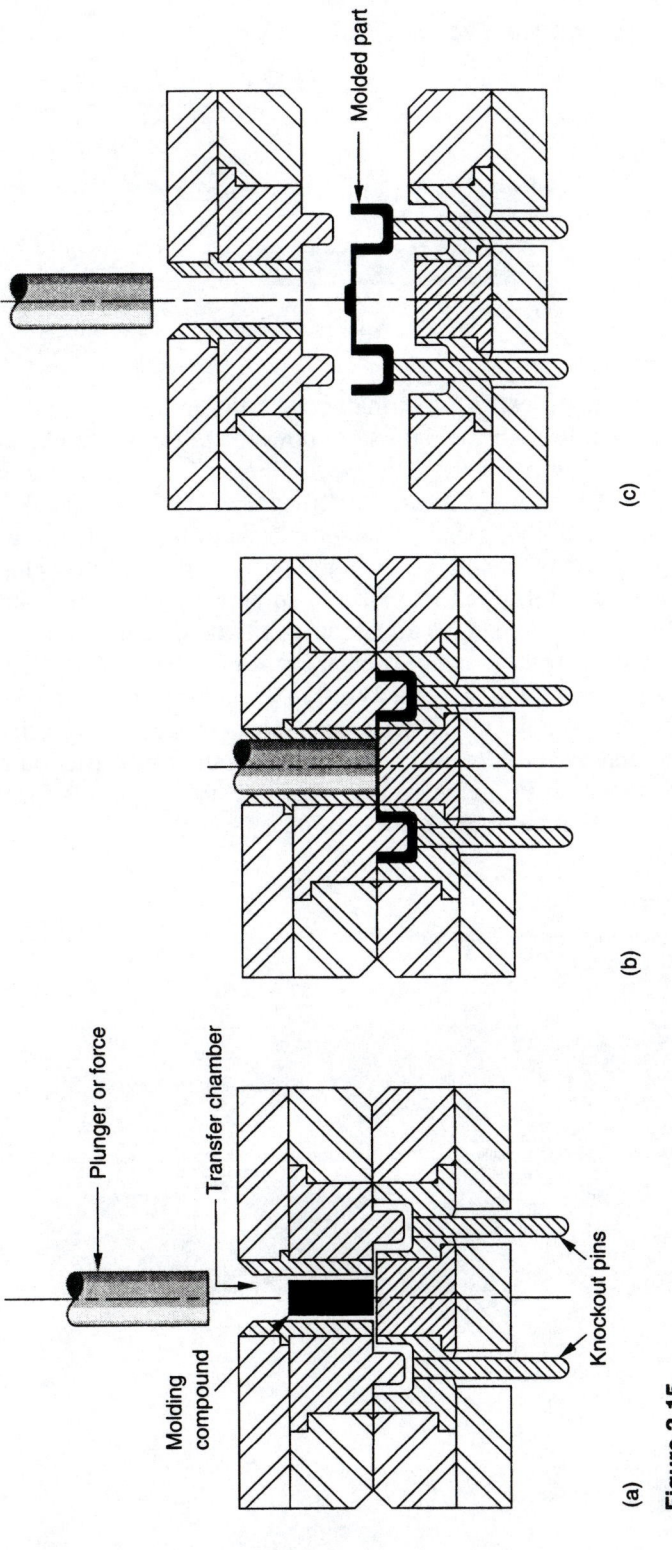

**Figure 3.15**
Transfer molding: (a) placement of molding compound, (b) transfer through gate system, (c) ejection of molded part
(Reproduced by permission. *Industrial Plastics: Theory and Application*, 2nd ed., by T. L. Richardson. Delmar Publishers, Inc., Albany, NY. Copyright 1989.)

**Figure 3.16**
Conventional pultrusion

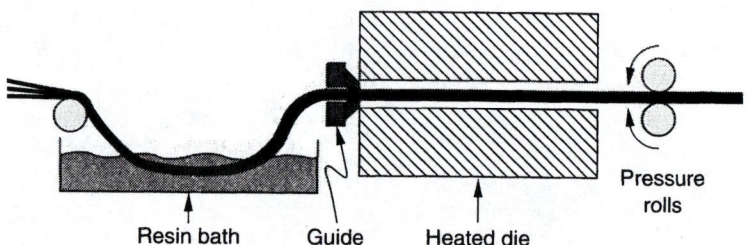

Resin bath          Guide          Heated die          Pressure rolls

metal wire drawing dies. In most instances, reinforcing material is soaked in the resin, then drawn through the die where polymerization takes place. The process is shown schematically in Figure 3.16. Glass fibers are the most common reinforcement, but aramid, graphite, carbon, and other fibers are also used.

Pultrusion is limited to simple shapes, except when accelerated curing is incorporated into the processing. For example, curved dies can be closed on the pultruded parts to accelerate curing without interfering with the continuous pulling. Skis are made in this manner, as are hammer and other tool handles.

Advanced pultrusion processing incorporates resin injection through ports near the entry to the die to provide thorough wetting of the fibers, as shown in Figure 3.17. By elimination of the resin dip tanks, moisture absorption problems are also reduced and by use of newer acid-free phenolic resins, friction and adhesion in the dies are reduced. Products such as flame-retardant panels for naval ships and building construction are made by pultrusion.

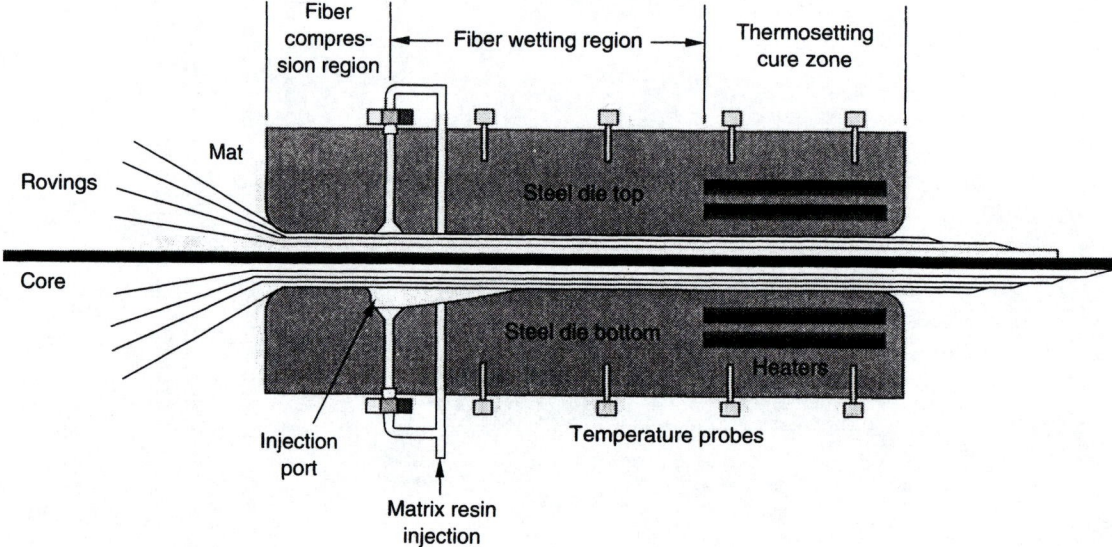

**Figure 3.17**
Advanced pultrusion processing

# 3.2.4 Casting Processes

**Casting** is an important method for shaping of thermosetting polymers such as phenolics, epoxies, and polyurethanes and for silicone elastomers. It is also practiced for a few thermoplastic materials containing **plastisols** (polymer solvents) that are cast as films and for parts where polymerization takes place in situ. In order to fill a mold without added pressure, however, few thermoplastics have sufficiently low viscosity to be cast to shape. Thermosets, on the other hand, are readily cast, particularly the unsaturated polyesters, epoxies, acrylics, phenolics, and polyurethanes.

Casting is most often associated with filling a shaped mold with liquid material, whether it is molten metal or a thermosetting polymer. However, there are a number of different casting methods that are practiced, such as film casting (which we have already discussed), dip casting, static casting, and centrifugal casting. One of the advantages of casting is the ability to make patterns that are simple or complex inexpensively. For example, many decorative items such as picture frames and furniture parts and even large foam blocks are made by casting. At the same time, tooling costs for small production quantities are minimized by selection of casting as the manufacturing process.

One of the most important applications of simple casting into a mold is the **encapsulation** of parts, such as electronic components for resistance to mechanical shock, thermal cycling, and protection against moisture or chemical attack. Either polyester or epoxy resins are used for this purpose, but they are modified with fillers, such as mica, to match the thermal expansion of components as closely as possible. *Vacuum-assisted casting*, depicted in Figure 3.18, is used mainly for small quantities or for very large parts where automation is impractical; transfer injection molding is, of course, preferred for high-production encapsulation.

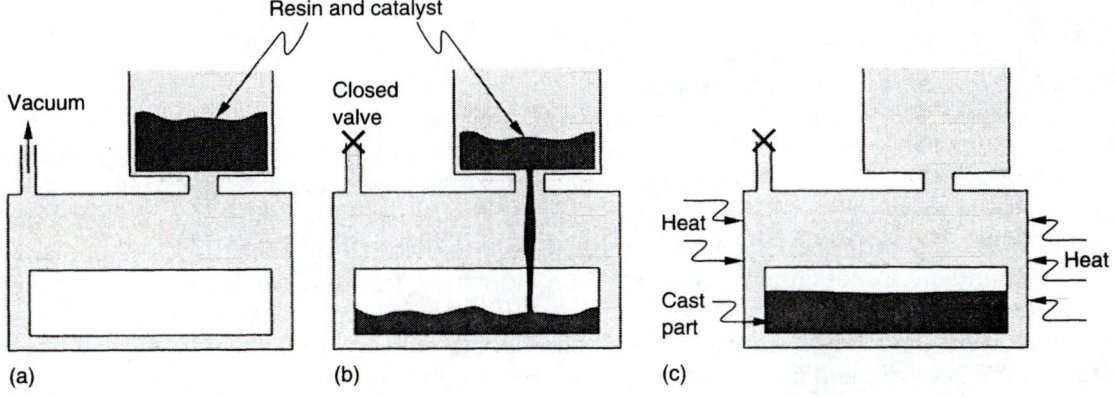

**Figure 3.18**
Vacuum-assisted casting: (a) evacuation of mold and resin mixing, (b) casting, (c) curing

**Dip casting** differs from dip coating because the form is removed from the final product; coatings, on the other hand, remain as part of the total product. Dip casting uses heated molds or forms that are lowered into highly plasticized polyvinyl chloride. The heat immediately fuses the plastisol closest to the form and voids are caused by evaporation of the plasticizers through the remaining plastisol. Thickness of the fused layer of course depends on the time and temperature, but the form can be removed from the plastisol when the desired thickness is achieved and placed in a heated conveyor to complete the fusion. The product is stripped from the form and trimmed. Industrial-grade rubber gloves, rain boots and rubbers, handle grips, and many other common items are made in this manner.

A variation of dip casting is *slush molding*, whereby plastisols are cast into a heated mold. When the desired wall thickness is achieved, excess plastisol is poured from the mold and the mold is placed into an oven to complete the fusion. Even dry thermoplastic powders have been used in the same way in a process called *static casting*, whereby the mold filled with powder is heated, causing fusion adjacent to the mold walls. Again, when the desired wall thickness is achieved, excess powder is poured from the mold and the mold is placed in an oven to complete fusion.

**Centrifugal casting** and **rotational casting** are modifications of simple casting of thermosetting materials or slush casting of plastisols. The difference is that the mold is rotated while the fusion of plastisols or polymerization of thermosets takes place. Centrifugal force then causes the material to take the shape of the mold. The difference between the two processes is that centrifugal casting involves only one axis of rotation, whereas rotational casting involves rotation on two axes. Many items such as balls, toys, industrial containers, bearings, and even furniture are produced by this method.

---

## Case Study 3.4

### Today's Phenolics

Phenolics, one of the earliest thermosetting materials, are based on the condensation reaction of phenol and formaldehyde to form a resin that is mixed with reinforcing materials, then transformed under heat and pressure to a highly cross-linked product. All phenolic molding compounds are reinforced, using glass fibers in products such as circuit breakers, carbon fibers in aircraft components, and silica fibers for thermal resistance in aerospace applications. The horizons for phenolics have been expanded in many cases because of improvements in processing techniques. At PF Industries, for example, the development of the advanced pultrusion processing (Figure 3.17), in combination with availability of new acid-free resin, has led to pultruded phenolic panels with minimum voids, high flexural strength (93,000 psi), and superior flame-retardant capability.

The heat resistance of phenolics has also been used to advantage by PF Industries in applications for the auto industry. Processing of complex shapes was made possible by **lost-core injection molding**, whereby hollow cores of a low melting point metal are made by blowing nitrogen and the molten metal into a core mold, then pouring excess metal away once a solid skin has formed. These cores are inserted into the injection mold cavity and the reinforced phenolic is injected around the core (holes in the core are in positions that are covered during phenolic injection). After ejection, the molded phenolic is placed in hot oil that melts out the hollow core, producing the desired cavity in the finished part. Hollow cores have increased the production rate several times simply because of the reduced time needed to melt out a hollow core in comparison to that for melting a solid one.

# Summary

Many methods are used for shaping polymer parts. In this chapter, we have separated them into those that are used mainly for thermoplastic polymers and those used mainly for thermosetting polymers. The most important properties for processing of thermoplastic materials are the viscosity and flow characteristics; the melt index is used as a guide for actual processing temperatures. In these materials, extrusion, injection molding, and thermoforming are the major production methods. Combinations of processes such as extrusion of sheet and thermoforming of that sheet are common in the industry. Miscellaneous processes for shaping thermoplastic polymers include blow molding, film casting, expansion processes to make foam, calendering, and laminating. Major processes for shaping thermosets are compression molding, transfer molding (which combines compression molding with injection molding), pultrusion, whereby thermosets are pulled through a die as curing takes place, and casting. In addition to conventional casting, dip casting and centrifugal or rotational casting are important processes for producing thermosetting materials.

# Terms to Remember

| | |
|---|---|
| blow molding | dip casting |
| blown film | drape molding |
| calendering | encapsulation |
| cast film | expandable styrene |
| casting | expanded foam |
| centrifugal casting | extrusion |
| compression molding | injection molding |
| crystallinity | laminating |
| denier | lost-core injection molding |

melt flow rate (MFR)

melt index (MI)

plastisol

processing

pultrusion

rotational casting

screw spinneret

thermoforming

thermoplastic

thermosetting

transfer molding

vacuum forming

viscosity

# Problems

1. Explain the importance of the melt index or melt flow rate in processing polymers.
2. Describe in your own words how plastic parts are extruded, giving examples of extruded plastic products.
3. Describe in your own words how plastic parts are injection molded, giving examples of injection-molded plastic products.
4. Describe in your own words how polyester can be made into a fabric or into a 2-liter bottle.
5. Make a list that describes different applications you know of for foams.
6. Compare the molding processes for phenolic to that of a multilayer printed wiring board made of prepreg. (*Hint:* See Chapter 8.)
7. Describe in your own words the transfer molding process.
8. Explain why it is more logical to study pultrusion as a polymer processing method rather than as a composite processing method.
9. Explain in your own words what the hollow core process is.
10. Make a list of products that you know of that are made by casting polymers.

# 4

# *Crystalline Ceramic Materials*

Ceramic materials are **inorganic**, **nonmetallic** materials that usually consist of metallic and nonmetallic elements bonded together primarily by **ionic** and/or **covalent bonds**. The chemical compositions of ceramic materials vary considerably from simple compounds to bonded mixtures of many complex phases.

The variation in properties of ceramic materials is large because of differences in bonding and structure. In general, ceramics are hard and brittle, with low toughness and essentially no ductility at room temperature. With some exceptions, ceramics are usually good electrical and thermal insulators because of the absence of conduction electrons. Ceramics normally have relatively high melting temperatures and are chemically stable in many hostile environments due to the stability of their strong bonds. Although we use ceramic materials for many applications, it is convenient to divide them into two groups, traditional ceramics and technical ceramics.

We make **traditional ceramics** from three basic components—clay, silica, and feldspar—that we mine from naturally occurring deposits. Clay is basically a hydrated aluminum silicate mineral with a very fine sized platelike structure. These platelets provide plasticity for wet forming as well as strength after drying and firing for many traditional ceramic products, such as bricks, tile, porcelains, sanitary ware, and more.

Silica ($SiO_2$) is also an important material for traditional ceramics. We use it in glass, whiteware ceramics, various bricks, and abrasives. Silica is both plentiful and widespread in the earth's crust and is one of the purest of the abundant minerals. The most common form of silica is crystalline quartz.

Feldspars, which are basically sodium potassium aluminum silicates, are also common minerals used in traditional ceramics. They act as fluxes in whiteware compositions and as a source of alkali and aluminum oxide in glasses, glazes, and enamels. Feldspars are inexpensive and are a source for water-insoluble alkalis.

**Modern**, or *technical*, **ceramics**, by contrast, typically consist of pure or nearly pure compounds such as magnesium oxide, aluminum oxide, barium titanate, silicon carbide, and silicon nitride. These compositions are most often synthesized by chemical reactions.

# 4.1 Properties of Ceramics

Ceramic applications, both traditional and technical, depend on the unique properties of the ceramic that, in turn, depend to a large extent on crystal structure. **Density**, for example, depends on the atomic weight of atoms present and on the atomic packing. Table 4.1 lists the densities of ceramic compounds.

**Table 4.1**
Densities of some ceramic compounds

| Compound | Specific density (g/cm³) |
|---|---|
| Low-density elements | |
| Boron carbide, $B_4C$ | 2.51 |
| Boron nitride, BN | 2.2 |
| Quartz, $SiO_2$ | 2.65 |
| Silicon carbide, SiC | 3.2 |
| Silicon nitride, $Si_3N_4$ | 3.2 |
| Intermediate-density elements | |
| Magnesium oxide, MgO | 3.65 |
| Aluminum oxide, $Al_2O_3$ | 3.99 |
| Titanium dioxide, $TiO_2$ | 4.25 |
| Barium titanate, $BaTiO_3$ | 6.0 |
| High-density elements | |
| Thorium oxide, $ThO_2$ | 9.8 |
| Hafnium oxide, $HfO_2$ | 10.1 |
| Tungsten carbide, WC | 15.7 |

# 4.1.1 Atomic Packing Factor

The **atomic packing factor** (APF) is a useful parameter for understanding crystalline ceramic materials. It is simply the volume fraction of atoms in a unit cell and can be derived from the atomic radius, if it is known, from the geometry of the unit cell, or from the density of the material. For a simple cubic structure, there are eight corner atoms that touch along the unit cell length, but each corner atom is shared by eight unit cells. Hence only one atom ($8 \times 1/8$) belongs to the unit cell. The volume of the atom is $(4/3)\pi r^3$, but the volume of the cell is $(2r)^3$. The packing factor is then $(4/3)\pi r^3 / (2r)^3$, or 0.5236. For more complicated geometries, such as the diamond structure, it is easier to use the relation

$$APF = \rho \times AN \times \frac{{}^4/_3 \pi r^3}{AW}$$

where $\rho$ is the density of diamond, 3.51 gm/cm³; AN is Avogadro's number, $6.023 \times 10^{23}$; $r$ is the radius of the carbon atom, $0.77 \times 10^{-8}$ cm; and AW is the atomic weight of carbon, 12.01. Thus, the APF of diamond is 0.34. For compounds, the crystal structure, the number of each atom species, and the unit cell dimensions must be known.

In general, compounds with low packing factors, like that for diamond, tend to have low expansion coefficients. In such cases, much of the expansion is absorbed by the open spaces in the structure so that the total expansion is minimized. Materials with high packing factors, that is, above 0.65, are close packed ionic or atomic structures and the expansion of each atom is additive throughout the structure. Figure 4.1 shows a plot of expansion coefficient versus temperature for various materials.

# 4.1.2 Thermal Conductivity

**Thermal conductivity** ($K$) is the rate of heat flow per unit area through a material. Ceramic materials exhibit a wide range of thermal conductivity behavior. The highest conductivities are observed in the least cluttered ceramic structures, that is, structures consisting of a single element, structures made up of elements of similar atomic weight, and structures with no extraneous atoms in solid solution. Diamond is a good example of a single-element ceramic structure with a very high thermal conductivity, more than double that of copper at room temperature. BeO, SiC, and $B_4C$ are examples of ceramic compounds made up of elements of similar atomic weight and size that have high thermal conductivity. Phonons, which conduct heat and electricity in ceramics much like conduction electrons do in metals, can move easily through these structures because the lattice scattering is small. In other materials such as $UO_2$ and $ThO_2$, where there is a large difference in the size and atomic weight of the ions, lattice scattering of phonons is larger and the conductivity is subsequently low. $UO_2$ and $ThO_2$ have less than one-tenth of the conductivity of BeO

**Figure 4.1**
Thermal expansion charac-
teristics of typical metals,
polymers, and polycrystalline
ceramics
(From D. W. Richerson,
*Modern Ceramic Engineer-
ing: Properties, Processing,
and Use in Design,* Marcel
Dekker, Inc., 1992.)

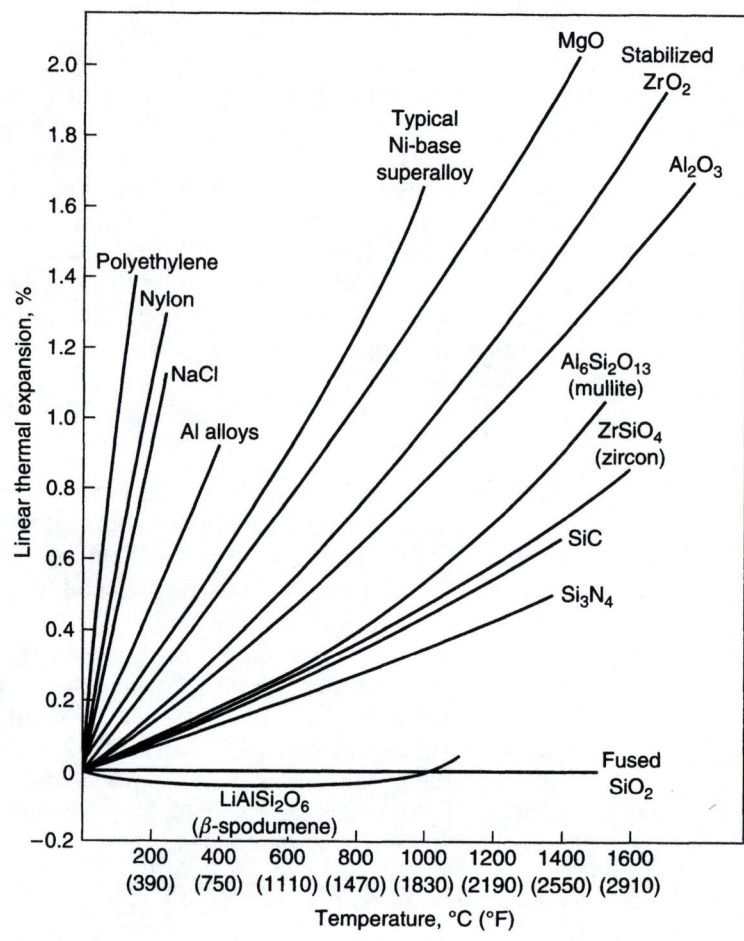

and SiC. Materials such as MgO and $Al_2O_3$ have intermediate values. Figure 4.2 shows a plot of conductivity versus atomic weight for several ceramic materials.

# 4.1.3   *Mechanical Strength*

As a class of materials, ceramics are **brittle** and the observed **tensile strength** varies over a large range, from values of several hundred psi to over a million psi for single-crystal **whiskers** of $Al_2O_3$ and SiC. (Whiskers are fine fibers that have large length-to-diameter ratios.) Of course, we must prepare such whiskers under carefully controlled conditions. Few ceramics in bulk form have tensile strengths above 25,000 psi. These materials also exhibit a large difference between their tensile and compressive strengths, the compressive strengths usually being about five to ten times higher than the tensile strengths. Most ceramics are hard and have low impact

**Figure 4.2**
Effect of cation atomic weight on the thermal conductivity of some ceramic oxides and carbides

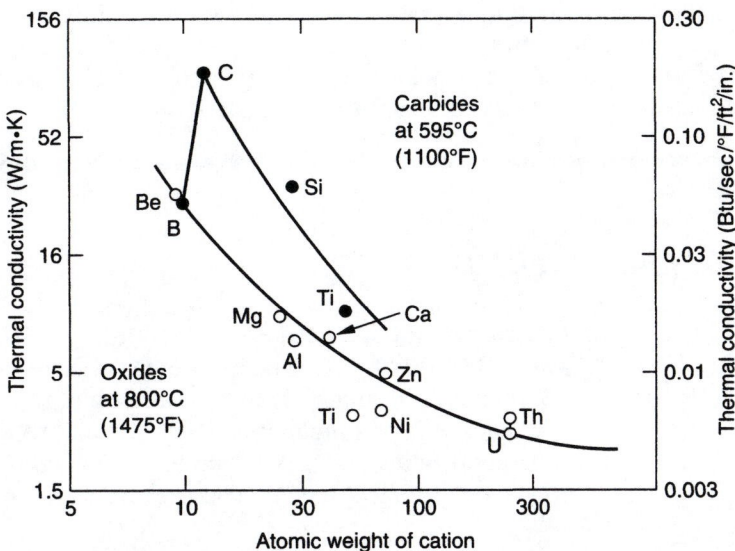

resistance because of their ionic and covalent bonding. The mechanical failure of ceramic materials occurs mainly from structural defects. The principal sources of fracture in ceramics are surface cracks produced during surface finishing, voids (porosity), inclusions, and large grains produced during processing. These play a critical role in determining the observed fracture strength of a ceramic material. In fully dense ceramics, that is, where there are no large pores, the strength is inversely related to the grain size:

$$\sigma \sim kd^{-1/2}$$

We recognize this relation, of course, as the Hall-Petch equation described in volume 1 on metals and alloys.

---

## Case Study 4.1

### Cutting Tools

The successful development of ceramic aluminum oxide cutting tools for cutting hardened tool steels and cast iron has been based largely on the achievement of controlled, fine-grained microstructure with minimal residual porosity. Prior to this development, material with larger grain size would fail unpredictably and often too soon to be of much practical value in production machine shops.

In order to synthesize these cutting tools, pure and fine particulate source material was needed to fabricate shaped cutting tools. The fine-grained structure achieved (1–2 micron grain diameter) raised the tensile strength to levels approaching 100,000 psi and minimized chipping of the tool edge to the dimensions of the size of the grains themselves. More recently, cutting tools based on silicon nitride have been developed that are even better for machining cast irons. These compositions are also dense, with fine grain size microstructure to avoid premature catastrophic failure.

## 4.1.4   Toughness

Ceramic materials, because of their combination of covalent and ionic bonding, have inherently low toughness. Research in recent years has focused on improving the toughness through use of fiber and whisker additives as well as by microstructure control. Fracture toughness testing has played a key role in this research, using a four-point bend test to determine $K_{1c}$ values in a manner similar to the fracture toughness testing of metals. The fracture toughness equation

$$K_{1c} = YS_f \sqrt{\pi a}$$

relates fracture toughness values to the fracture stress, $S_f$, and the largest flaw size, $a$. $Y$ is a dimensionless constant equal to about one. In this relation, $K_{1c}$ is measured in MPa m$^{1/2}$ (ksi in.$^{1/2}$).

Values for monolithic ceramics are generally quite low, ranging from 2 to about 6 MPa m$^{1/2}$. We will describe how to improve the toughness of these materials by the composites approach in Chapter 7.

## 4.2   Thermal Shock Behavior

The susceptibility of ceramic materials to thermal stresses and **thermal shock** failure is one of the main factors limiting their more widespread application. Although many ceramics are capable of extensive useful life at elevated temperatures, failure can occur during heating and cooling. Under these conditions, the **temperature gradient** between the interior and exterior surfaces can develop large stresses in the material. The magnitude of the gradient is dependent on the thermal conductivity of the ceramic and the rate of heating or cooling. The mechanical stress that develops is given by the relation

$$S = \frac{E\alpha\Delta T}{1 - \mu}$$

where $S$ is the resulting stress if the body *is unconstrained*, $E$ is the elastic modulus, $\alpha$ is the coefficient of linear thermal expansion, $\Delta T$ is the temperature gradient in the ceramic body, and $\mu$ is Poisson's ratio (0.2 to 0.3 are common for ceramics).

We don't normally encounter failure by thermal shock in metallic materials because such stresses are relieved by plastic deformation. The brittle behavior of ceramics precludes this relief mechanism at low temperatures.

---

### Sample Problem 4.1

High-quality aluminum oxide has the following properties:

$$E = 50 \times 10^6 \text{ psi}$$
$$\alpha = 9 \times 10^{-6}/°C$$
$$\mu = 0.25$$

What type of stress is developed on the surface of the ceramic upon heating or cooling? What is the amount of stress if the gradient in temperature is 100°C?

### Solution

The surface will be under compression upon heating because it will expand faster than the interior and under tension upon cooling because it will contract faster than the interior.

$$S = \frac{(50 \times 10^6) \times (9 \times 10^{-6}) \times 100}{1 - 0.25}$$
$$= 60,000 \text{ psi}$$

This stress can easily exceed the ultimate tensile strength, fracturing the ceramic catastrophically.

---

Because most dense, high-quality ceramics that are considered for advanced structural applications have high elastic moduli, for example, in the range 45–65 × $10^6$ psi, emphasis is placed on selecting those materials with low coefficients of expansion for stringent structural application.

---

## Case Study 4.2

### Ceramics for Advanced Turbine Engines

In the early 1970s, the U.S. government provided funding support for an extensive effort that would lead to the development of a prototype turbine engine based on ceramic hot-stage components capable of operating at 2500°F for extensive periods of time. Material selection proved to be a very important part of the overall program. Strength, fracture toughness, raw material cost, fabricability, oxidation resistance, and thermal shock resistance were properties and characteristics that it was hoped could be embodied in a single ceramic material. The selection of silicon nitride, $Si_3N_4$, as the primary material was based

**Figure 4.3**
Photograph of silicon nitride turbine rotors
(From D. W. Richerson, *Modern Ceramic Engineering: Properties, Processing, and Use in Design,* Marcel Dekker, Inc., 1992.)

to a large extent on its low coefficient of thermal expansion (about one-third that of aluminum oxide) that enables it to survive steep temperature gradients of at least three times that for aluminum oxide, which was the leading technical ceramic at the time. Although the target temperature of 2500°F has yet to be reached for extended time periods, the effort continues toward developing prototype engines suitable for automobile vehicular use, and silicon nitride is still the prime material today. Figure 4.3 shows a photograph of silicon nitride turbine rotors manufactured for testing in experimental turbine engine assemblies.

# 4.3  Crystal Structure

Most ceramic materials are crystalline compounds formed by reaction between the nonmetallic and metallic elements of the periodic table. The bonding that holds these compounds together is ionic for the most part and results from the coulombic force of attraction between negative ions (anions) and positive ions (cations). The cations are formed by the loss of valence electrons from metallic elements. Nonmetallic elements become the anions during compound formation. For these materials, the structure is in large part determined on the basis of how positive and negative ions can be packed to maximize electrostatic attractive forces and minimize electrostatic repulsion.

# *4.3.1* *Pauling's Rules and Coordination Number*

We can understand most ceramic **crystal structures** because of general observations credited to Linus Pauling, an early crystallographer. Although there are five general statements that we call Pauling's rules, we will concentrate only on the first. This rule states that a coordination polyhedron of anions surrounds each cation and that the **coordination number**, that is, the number of anions around the cation, is determined by the ratio of the radii of the two atoms.

Coordination numbers are greatest for the largest values of radius ratios, as depicted in Figure 4.4. The critical ratios presented in this figure are followed in many ceramic crystal structures. For example, oxides such as MgO, NiO, CaO, and FeO have a radius ratio of cation to anion that exceeds 0.414 but is less than 0.732, so we find the coordination number for each of these compounds is 6. The unit cell typical of these materials is the NaCl structure shown in Figure 4.5. Here we can see that the $Na^{+1}$ ion in the center of the cubic unit cell has six $Cl^{-1}$ ions, located in the centers of the cube faces, surrounding it.

Another of Pauling's rules stipulates that the sum of positive and negative ion charges in the cell must equal zero to preserve stability and minimum energy requirements. This rule also is satisfied as we add up sodium ions in the unit cell (a total of four) and we get the same number for chloride ions. To arrive at these totals

**Figure 4.4**
The coordination number
and the radius ratio

| Coordination number | Location of interstitial | Radius ratio | Representation |
|---|---|---|---|
| 2 | Linear | 0–0.155 |  |
| 3 | Corners of triangle | 0.155–0.225 |  |
| 4 | Corners of tetrahedron | 0.225–0.414 | 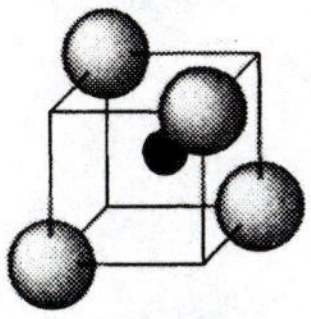 |
| 6 | Corners of octahedron | 0.414–0.732 |  |
| 8 | Corners of cube | 0.732–1.000 | 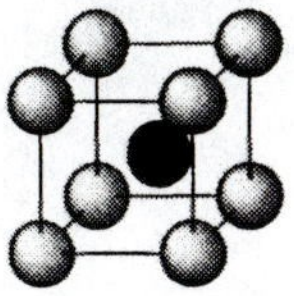 |

**Figure 4.5**
Crystal structure of sodium chloride: (a) schematic of open crystal structure showing the location of positive and negative chloride ions, (b) close-packed structure showing contact between unlike ions

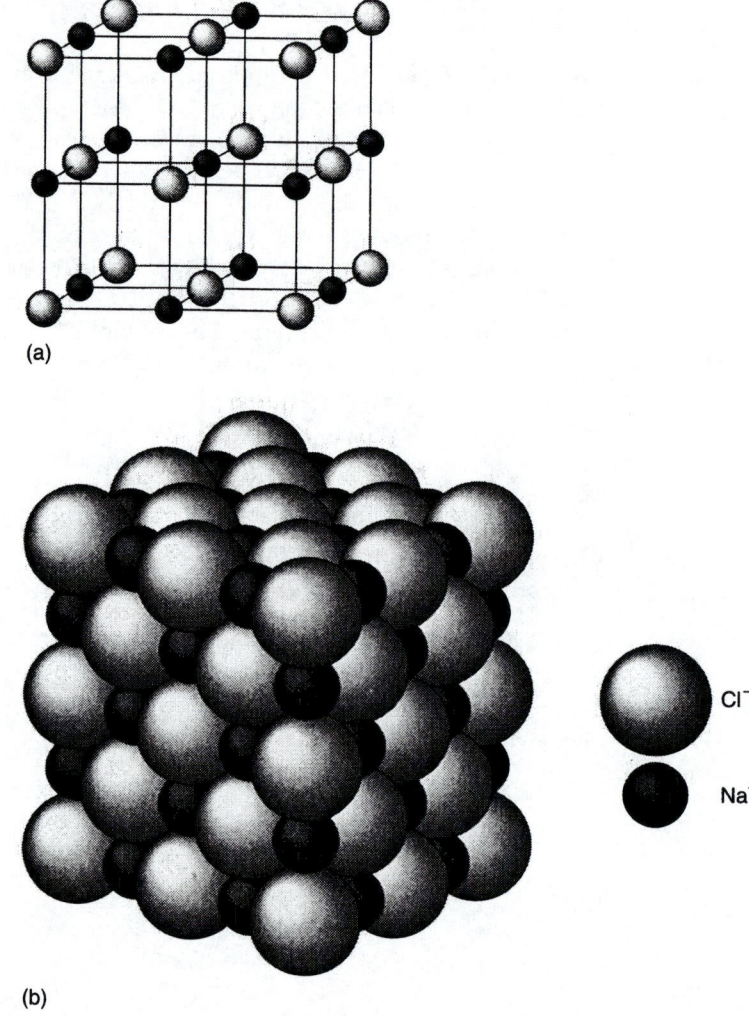

(a)

(b)

Cl⁻

Na⁺

we must consider that the ions on the corners, edges, and cube faces of the unit cell are shared by adjoining unit cells. In counting $Na^+$ ions, there is one located in the center of the cell, which is not shared by any other cell, and there are 12, each of which is located at the center of one of the 12 cube edges. Each one of these ions is shared by 4 other adjoining unit cells, so there are 12/4, or 3, $Na^+$ ions in our cell from these locations, which gives us a total of 4 $Na^+$ ions. Similarly, for chloride ions located at the 8 corners of the cube, we find that each one of these is shared by 8 other adjoining cells. Therefore, we have 8/8, or 1, $Cl^-$ ion for the cell from this location, plus we have $Cl^-$ ions located at the center of each of the 6 cube faces. Each one of these is shared by one other adjoining cell, so we have a total of 3 for the cell from this location. The total of chloride ions from the cube corner and cube face locations is, then, 1 plus 3, or 4, which neutralizes the 4 sodium ions. The most important ceramic compound displaying this structure is MgO, where $Mg^{+2}$ replaces $Na^{+1}$ and $O^{-2}$ replaces $Cl^{-1}$. The net charge adds up to zero, and the radius is between the limits 0.41 and 0.73 so that the coordination number of 6 is maintained.

# *4.4 Technical Ceramic Compounds* ___

## *4.4.1 Compounds with the NaCl Structure*

MgO is a good electrical insulating material at elevated temperatures and is often used as the high-temperature insulator encircling the heating elements used in electric stoves and ovens. It is an important refractory in steel plant furnaces as well as a constituent in many glass compositions. When fabricated as a pure, dense material with essentially zero porosity, it is optically transparent and has been considered as a useful infrared transmitter in various systems.

Other very useful compounds crystallizing with the sodium chloride structure include refractory carbides and nitrides of titanium, zirconium, and so on, that are related to tungsten carbide and form the basis of the carbide tool industry. These are unusual in that the carbon atom replaces the sodium ion because it is so small, and the metal atom, titanium in the case of TiC, replaces the chloride ion in the sodium chloride structure of Figure 4.5. In fact, the radius ratio of carbon to titanium is greater than 0.41 but less than 0.732 and the coordination number of six is observed even though the bonding is not ionic. These materials have very high hardness, high melting points, well in excess of the metals they are derived from, and are good electrical conductors as well. These characteristics are related because the hardness is derived in part by the covalent bonding; however, carbon has only four valence electrons that must be shared with six titanium nearest neighbors. We believe the valence electrons resonate locally to provide the covalent bonds, leaving two titanium valence electrons free to provide the long-range electron mobility needed for a good conductor.

The nitrides behave in a similar fashion because the nitrogen atom is small enough to fit into the position occupied by carbon. With insufficient electrons needed for direct covalent bonding, resonance provides the mechanism for overall bonding and these compounds are also good electrical conductors. Titanium nitride in particular is very hard and in recent years has been applied as a coating on tools and drill bits to prolong life by reducing wear.

## *4.4.2 Compounds with the Fluorite Structure*

Many properties exhibited by ceramic compounds are determined by their crystal structure. Let's consider the cases of uranium dioxide, $UO_2$, and zirconia, $ZrO_2$, which have the same structure as calcium fluorite, $CaF_2$, shown in Figure 4.6. In this

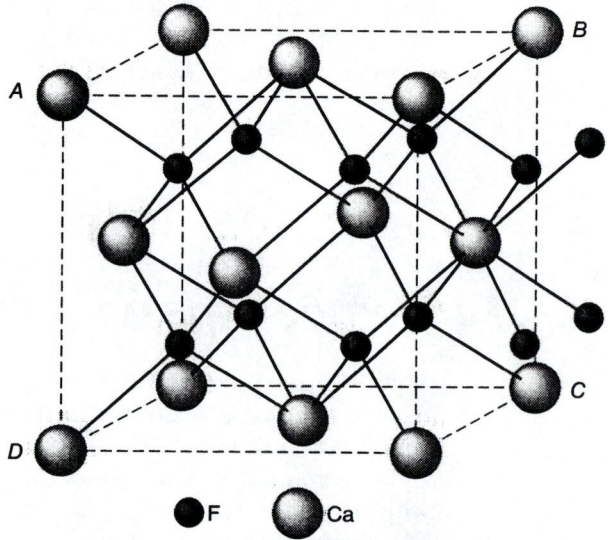

**Figure 4.6**
Fluorite structure
(From W. D. Kingery, H. K. Bowen, and D. R. Uhlmann, *Introduction to Ceramics,* 2nd ed., John Wiley & Sons, 1976.)

structure, eight fluoride ions are located within the unit cell comprising a face-centered cubic (fcc) array of calcium ions. The radius of $Ca^{+2}$ is 0.106 nm and that of the $F^{-1}$ is 0.133 nm, a radius ratio that requires the coordination number for the $Ca^{+2}$ ions to be eight, in agreement with the crystal structure in Figure 4.6. (Note that the atom sizes are reversed in this figure for easier visualization). The utility of the $UO_2$ and $ZrO_2$ structures is more apparent if we cut the unit cell to examine the plane defined by the atoms *A, B, C,* and *D*. This plane is shown in Figure 4.7. Both structures display a sizable hole in the center of the unit cell.

In the case of $UO_2$, which is used as a nuclear fuel element, the hole is approximately 0.21 nm, the size of the helium atom, which is a nuclear fission degradation product. Therefore, the crystal structure can tolerate the formation of a certain amount of fission products like He without fracturing and requiring premature replacement.

In the case of $ZrO_2$, the holes in adjacent unit cells provide a path for $O^{-2}$ ion diffusion at elevated temperatures. Because of this fact, it is used as an oxygen sensor on modern automobiles. The diffusion of oxygen ions is a function of oxygen

**Figure 4.7**
The (110) plane of the $UO_2$ and $ZrO_2$ crystal structure

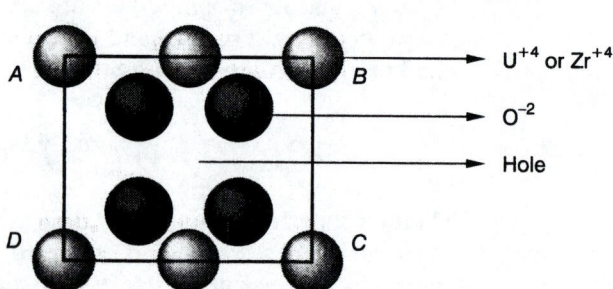

$U^{+4}$ or $Zr^{+4}$

$O^{-2}$

Hole

concentration in exhaust gases and is measured as an electric current that is then used to automatically monitor the fuel-air mixture to minimize pollution and improve fuel efficiency.

# 4.4.3  *Compounds with the Perovskite Structure*

The crystal structure of $CaTiO_3$, shown in Figure 4.8, is known as the **perovskite** structure. It is best visualized as a face-centered cubic lattice with calcium ions at the corners and oxygen ions at the face centers plus a titanium ion in the cube center. Perovskite is not important technically, but when we consider substitution of barium for calcium to form barium titanate, we find a very important technical ceramic. The barium substitution causes a tetragonal distortion of the unit cell because the titanium ion is too large for the center of the cube. As a result, the $Ti^{+4}$ ion shifts in one of the six coordination directions and the four $O^{-2}$ ions shift in the opposite direction, as shown in Figure 4.9. A permanent dipole moment results because the centers of these charges (between $Ti^{+4}$ and surrounding $O^{-2}$ ions) are displaced with respect to one another. This dipole moment is large enough that it provides $BaTiO_3$ with one of the highest dielectric constants ever observed, that is, in excess of 1000. In practical terms, a 3/8-in. diameter $BaTiO_3$ capacitor replaces a mica capacitor that is 15 in. in diameter for radios, television sets, and the like.

**Figure 4.8**
Perovskite structure (idealized)
(From W. D. Kingery, H. K.
Bowen, and D. R. Uhlmann,
*Introduction to Ceramics,* 2nd
ed., John Wiley & Sons, 1976.)

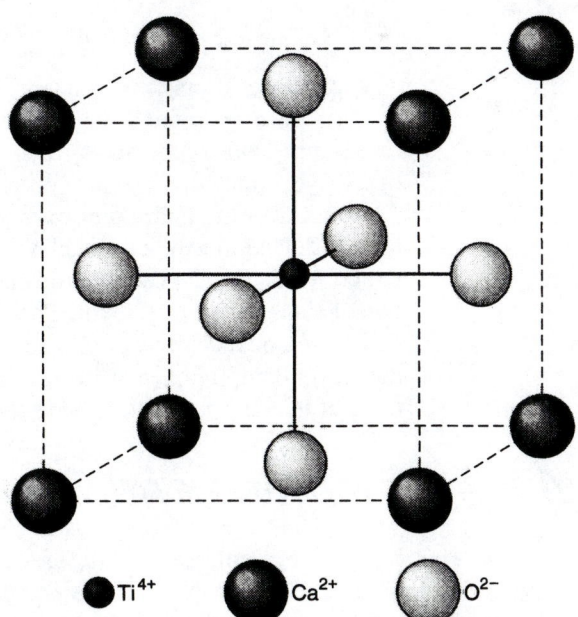

$Ti^{4+}$    $Ca^{2+}$    $O^{2-}$

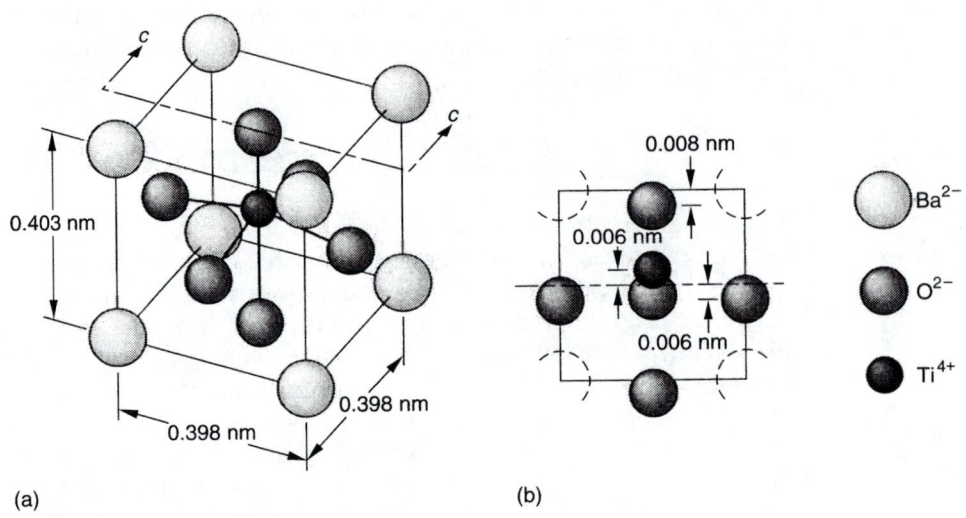

**Figure 4.9**
Barium titanate crystal structure: (a) tetragonal structure, (b) section *c-c* showing atomic shift that creates dipole
(From W. F. Smith, *Principles of Materials Science and Engineering*, 2nd ed., McGraw-Hill Publishing Company, 1990.)

# 4.5    Traditional Ceramic Compounds

## 4.5.1    Clays

The structure of **clay** minerals is typified by that of kaolinite, shown in Figure 4.10. Kaolinite is a hydrated aluminosilicate with the formula $Al_2Si_2O_5(OH)_4$, whose structure is based on combinations of an $AlO(OH)_2$ layer and a layer of silica $SiO_4$ tetrahedra. Oxygen ions in the figure projecting up from the silica tetrahedra are connected and built into the aluminum oxyhydroxide layer. Because no primary bonding exists beyond the connected layers, clay particles have a two-dimensional, platelike physical structure. Cleavage occurs readily across the layers, consequently the average particle size of clays is also very small. These structural characteristics give clay excellent plasticity when mixed with water, which is a very important processing feature for most of the traditional ceramics products industries. Water films between the platelike surfaces of the clay particles provide the vehicle for plastic behavior.

## 4.5.2    Diamond and Graphite Forms of Carbon

Several covalent bonded materials also display the characteristics of ceramic compounds. Chief among these is the **diamond** form of carbon where each carbon atom with four valence electrons attains a stable eight-electron configuration by bonding

(a)

(b)

**Figure 4.10**
Structure of kaolinite clay: (a) layered crystal structure of kaolinite
clay with Si-O tetrahedrons on the bottom half of the layer and Al-O,
OH octahedrons on the top half, (b) scanning electron micrograph
of kaolinite platelet structure (3000×)
(Part (a) from W. D. Kingery, H. K. Bowen, and D. R. Uhlmann,
*Introduction to Ceramics,* 2nd ed., John Wiley & Sons, 1976.)

with four other carbon atoms in tetrahedral coordination. The diamond crystal
structure shown in Figure 4.11 is fcc with tetrahedral coordination. The high hardness, strength, and electrical insulation properties of diamond are well known and
result from the very strong directed covalent bonds.

The other common crystalline form of carbon is **graphite**, shown in Figure
4.12, which has a layer structure in which carbon atoms are held together by strongly

**Figure 4.11**
Crystal structure of diamond
(From W. D. Kingery, H. K.
Bowen, and D. R. Uhlmann,
*Introduction to Ceramics,* 2nd
ed., John Wiley & Sons, 1976.)

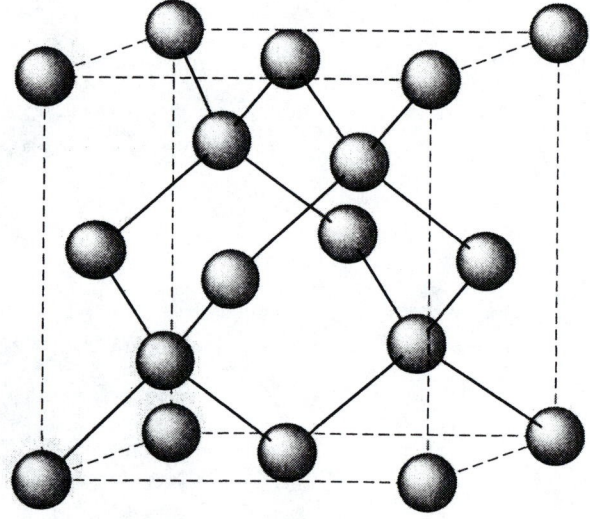

directed covalent bonds in a hexagonal array. By contrast, the bonds between layers
are weak, electrostatic ones and graphite, like clay, is essentially a two-dimensional
structure, with particles easily sliding over one another.

Each carbon atom in the hexagonal plane is covalently bonded to only three
nearest neighbors. The fourth valence electron actually resonates inside the hexagonal ring from one carbon atom to another as an added covalent bond with mobility—somewhat like the resonance that occurs in the structure of the refractory carbides mentioned earlier. The mobile electrons absorb light energy, giving graphite
its dark black color and making it a good electrical conductor, unlike the diamond

**Figure 4.12**
Graphite structure
(From W. D. Kingery, H. K.
Bowen, and D. R. Uhlmann,
*Introduction to Ceramics,*
2nd ed., John Wiley & Sons,
1976.)

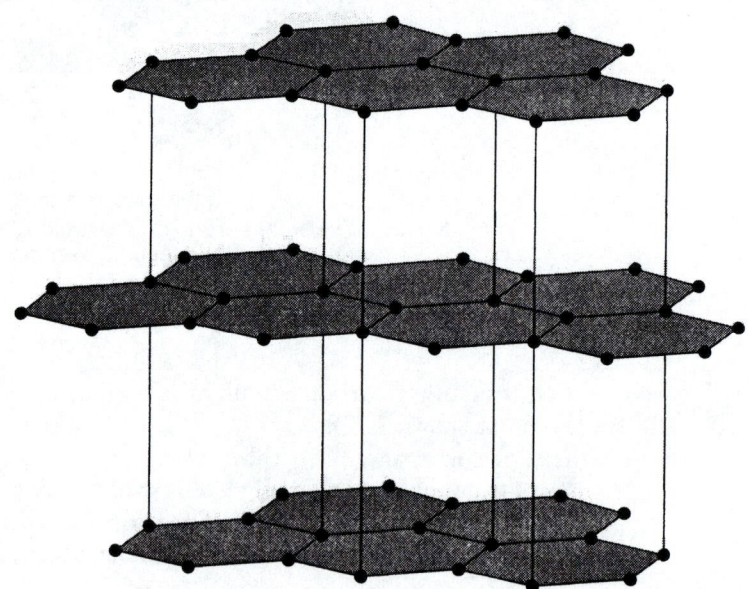

**Figure 4.13**
A subcell of magnetite

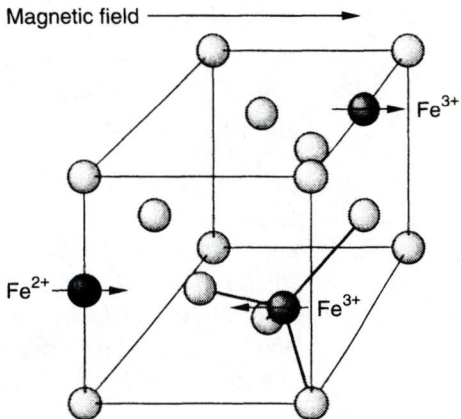

Magnetic field ⟶

$Fe^{3+}$

$Fe^{2+}$

$Fe^{3+}$

form. We find extensive use of graphite in high-temperature electrodes, heating elements, and molds for glass-forming operations.

## 4.5.3 Ferrites

Magnetic ceramics are based on the structure of magnetite, $Fe_3O_4$. Magnetite actually contains both ferrous ions ($Fe^{2+}$), and ferric ions ($Fe^{3+}$). We can therefore write the formula as $Fe^{2+}Fe_2^{3+}O_4^{2-}$, which demonstrates the 2:1 ratio of ferric to ferrous ions. This compound crystallizes into a spinel structure that is based on an fcc arrangement of oxygen ions, with iron ions occupying selected interstitial sites.

Rather than show the entire spinel unit cell, it is useful to describe one of the eight subcells that make up the structure, as illustrated in Figure 4.13. The oxygen ions are in the fcc positions of the subcell, and there are a total of four oxygen ions. One $Fe^{2+}$ ion and one $Fe^{3+}$ ion occupy octahedral positions at the center of the cube edge, while the second $Fe^{3+}$ ion occupies a tetrahedral position inside the cell.

The ions in octahedral sites align with the magnetic field, but ions in tetrahedral sites oppose the field. Equal distribution of ferric ions in these sites cancels any net magnetization, but all ferrous ions are in octahedral sites, giving a net magnetization. We call this phenomenon **ferrimagnetism** and many useful ferrite ceramics are produced. Perhaps the best known application of these is the ceramic magnets found in many homes (just look at almost any refrigerator door!).

## Summary

Ceramics are inorganic, nonmetallic materials consisting of both nonmetallic atoms and metallic atoms bonded together primarily by ionic bonds. Some ceramics, however, do have covalent bonds. As a consequence, the chemical compositions and structures of these compounds are varied, with significant property differences.

Density, expansion, and thermal conductivity depend on atomic weight and structure characteristics of the compounds. Ceramics include both traditional ceramics and modern, or technical, ceramics, depending on raw material sources and the applications of products.

Clay, silica, and feldspar minerals are major constituents of traditional ceramics that include bricks, porcelains, tile, piping, and more. Modern ceramics consist largely of specially processed compounds for particular applications, for example, barium titanate for its high dielectric constant, zirconium oxide as an oxygen sensor, silicon nitride as a turbine material, and magnetic ceramics based on iron oxide.

In most instances, ceramics are hard, brittle, good electrical insulators, and display a range of thermal conductivity and thermal expansion values depending on structure and chemical composition.

# Terms to Remember

| | |
|---|---|
| atomic packing factor | ionic |
| brittle | modern ceramics |
| clay | nonmetallic |
| coordination number | perovskite |
| covalent bond | temperature gradient |
| crystal structure | tensile strength |
| density | thermal conductivity |
| diamond | thermal expansion |
| ferrimagnetism | thermal shock |
| graphite | traditional ceramics |
| inorganic | whiskers |

# Problems

1. Define the term *ceramic*.
2. Discuss the difference between ionic and covalent bonding in ceramic compounds.
3. Explain what is meant by the terms *traditional ceramics* and *modern*, or *technical, ceramics*.
4. Account for the importance of clay in traditional ceramics.
5. What is the atomic packing factor and how is it related to the coefficient of thermal expansion?
6. Explain the variations in thermal conductivity observed for low-density and high-density ceramic oxide compounds.

7. Thermal shock sensitivity can preclude the use of some ceramics in thermal cycling applications. Explain why this is so.
8. Explain the relationship between coordination number and ionic radius ratio.
9. Account for the importance of the fluorite crystal structure in a ceramic compound used as an oxygen sensor.
10. Account for the high dielectric constant displayed by the compound barium titanate.
11. What are the significant differences between graphite and diamond?
12. What is ferrimagnetism?

# 5

# Glass

Glasses have special properties not found in other engineering materials. The combination of **transparency** and hardness at room temperature coupled with good corrosion resistance and sufficient strength make glasses indispensable for such applications as windows, beverage containers, laboratory ware, and more. In the lighting industry, glass is essential for various types of lamps because of its insulating properties and ability to provide a vacuum-tight enclosure as well as its transparency. In electronic applications, glass provides high-strength vacuum envelopes for electronic tubes along with insulation of electrical connectors. The high chemical resistance of glass makes it useful for laboratory apparatus and for corrosion-resistant liners for pipes and reaction vessels in the chemical industry. Glass-base enamels are useful in home applications such as stoves, washers, and dryers, and glass-base glazes form the exterior surfaces of most porcelain and chinaware as well as wall tile.

A *glass* is a ceramic material made from inorganic materials at high temperatures. We distinguish glasses from other ceramics because they are **vitreous**, or **amorphous**, that is, they are not crystalline. Glass, in fact, is defined as an inorganic material that has been cooled to a solid state without crystallization. The amorphous structure of glass is more like that of a liquid with only short-range order of the atomic bonds.

# 5.1  *Glass Structure*

We link the structure of glass to that of a liquid; however, the mechanical properties of glass show that atoms must be linked together to form an expanded, continuous three-dimensional network. In this network, the immediate environment of each atom is the same as or similar to that in a crystal, but there is no long-distance regularity. The distinction between a glass and a crystal has been characterized by Zachariasen using the two-dimensional example shown in Figure 5.1.

In the figure, there is no regularity in the lattice of the glass but the type of bonding and coordination is essentially the same in the two cases illustrated. In fact, the density of a pure silica glass ($SiO_2$) is 2.20 g/cm$^3$ and that for crystalline $SiO_2$ in the form of cristobalite (diamond structure) is 2.35 g/cm$^3$. These values suggest that the packing factors of these two materials are not far apart, a further illustration of similar coordination.

Glass melts gradually over a range of temperatures because of its amorphous structure. Since all the atomic bonds are not the same, the energy required to break an atom free from the network is different from atom to atom. When we increase temperature, therefore, atoms gradually become free, so that **melting** is a continuous rather than abrupt phenomenon, as shown in Figure 5.2. This figure compares cooling curves for both a glass and a crystal against specific volume (which is the reciprocal of density). A pure liquid that forms a crystalline solid upon cooling will solidify at a specific melting temperature with a large decrease in specific volume, as indicated by $T_m$ in Figure 5.2. In contrast, a liquid that forms a glass upon cooling

**Figure 5.1**
Planar structure of glass:
(a) regular crystalline lattice,
(b) irregular glassy network
(From W. D. Kingery, H. K. Bowen, and D. R. Uhlmann, *Introduction to Ceramics,* 2nd ed., John Wiley & Sons, 1976.)

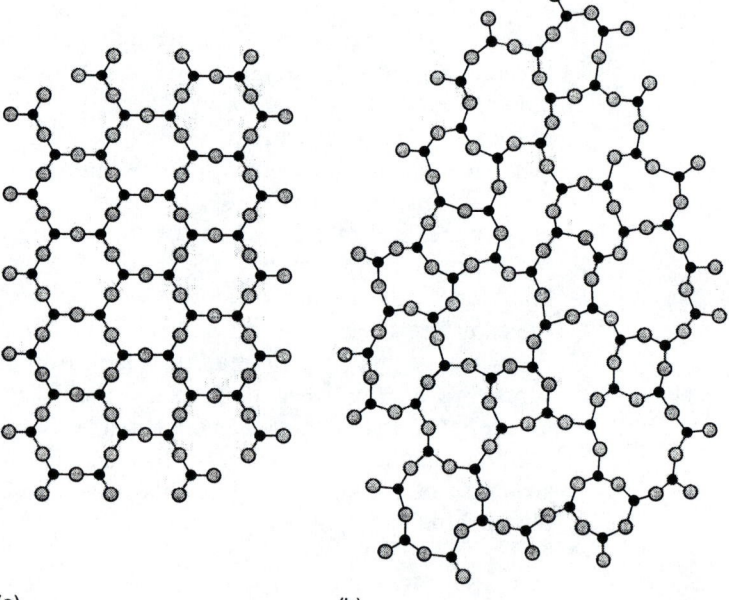

(a)                    (b)

does not crystallize but becomes more viscous as its temperature is lowered. It transforms from a very viscous fluid to a rigid, **brittle** state known as the *glassy state*. This transformation occurs in a narrow temperature range where the slope of the specific volume-versus-temperature curve is markedly decreased. The point of intersection of the two slopes of this curve defines a transformation point called the glass **transition temperature**, $T_g$. Below this temperature, glasses are, for all practical purposes, completely brittle. Each glass composition basically has its own transition temperature, however, this temperature does vary with cooling rate.

The conditions necessary for the formation of a glass are very specific. For example, a glass is stable only if its energy is nearly the same as that of the corresponding crystalline structure and if it is formed under conditions that prevent crystallization. In the vast majority of materials, however, the energy difference between a vitreous and a crystalline state is so large that glass can form only from exceptionally fast cooling rates, that is, $10^6$ degrees/sec. Amorphous metal particles are made by "splat cooling" at such rates, and in fabricated form they have unusual properties.

# 5.2 Glass Formation

Zachariasen has summarized four conditions that must be satisfied before an oxide glass of composition $A_mO_n$ is to be expected.

1. No oxygen atom is bound to more than two A atoms. This rule singles out those oxides that develop network structures.
2. The coordination number of oxygen around A must be low, usually three or four. This rule singles out those oxides with strong bonds.

**Figure 5.2**
Solidification of crystalline and glassy (vitreous) materials ($T_g$ is the glass transition temperature of the glassy material and $T_m$ is the melting point of the crystalline material.)

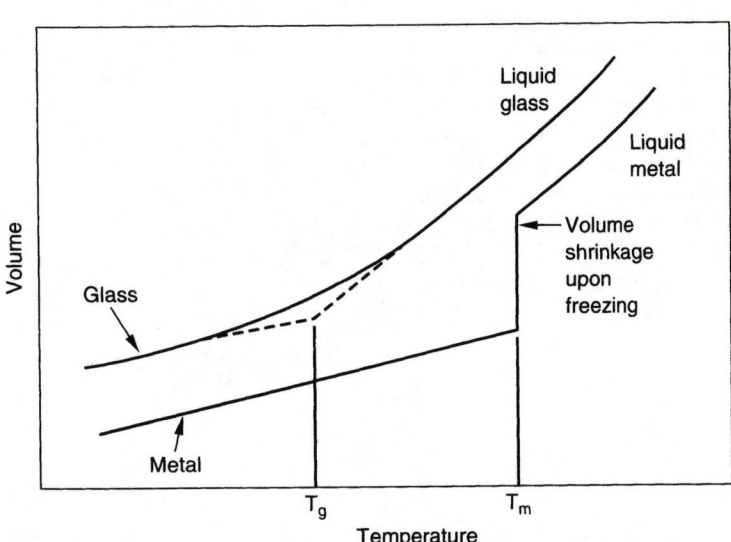

3. Oxygen polyhedra share only corners, not faces or edges.

4. At least three corners in each oxygen polyhedron must be shared. This rule ensures a three-dimensional network, which we sometimes refer to as polymerization of glass.

Relatively few oxides meet all of the Zachariasen rules. No elements with valence of one or two can form a network with oxygen, and only boron with valence of three is small enough to satisfy them. The group VI elements silicon and germanium and group V elements phosphorus and arsenic actually form glassy structures, but other elements from these groups do not.

In addition to the oxides that are **glass formers**, several oxides can enter an existing network to a limited extent. Alumina, $Al_2O_3$, is the most common example, because aluminum ions are small and can substitute for some of the silicon ions in the tetrahedral positions. It is probable that any network of tetrahedrally coordinated units can also contain a limited number of octahedral units. Thus other ions that behave similarly to aluminum, for example, zirconium, titanium, vanadium, antimony, and lead, can become part of a **glass network**.

Other oxides that are used extensively in glasses include $Na_2O$, $K_2O$, $CaO$, $BaO$, $MgO$, $PbO$, and the rare earth oxides. These are known as **glass modifiers** because the metallic element does not enter the network. In effect, these modifiers depolymerize the structure. Figure 5.3 compares the structure of a soda-lime glass with that of an alumina-silica glass. Note that the small size $Al^{3+}$ ions enter the network, but the $Na^+$ ions do not. Modifiers increase the workability, that is, the flow properties of a glass, because of the depolymerizing effect, but the extent of addition is limited. Network modifiers increase the ionic character of the glass and promote devitrification (crystallization).

**Figure 5.3**
Network modified glass (note that $Na^+$ ions do not form part of the network): (a) soda-lime glass, (b) alumina-silica glass (note that $Al^{3+}$ ions do form part of the network) (From W. F. Smith, *Principles of Materials Science and Engineering,* 2nd ed., McGraw-Hill Publishing Company, 1990.)

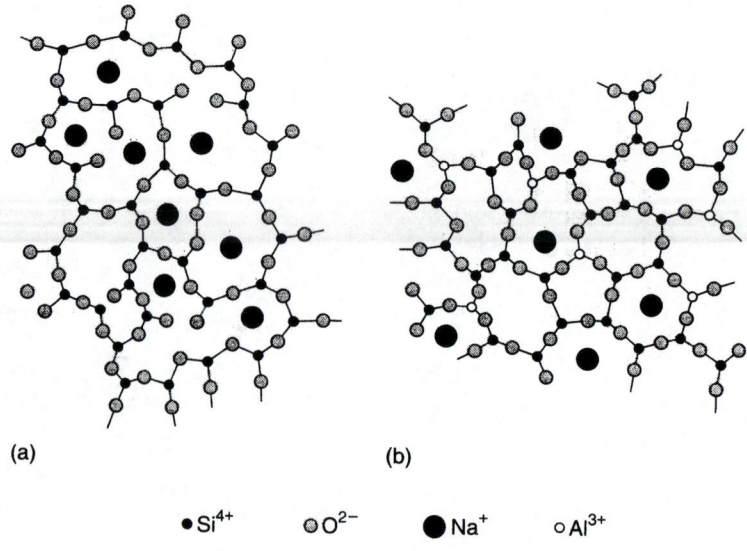

(a)  (b)

● $Si^{4+}$  ◎ $O^{2-}$  ● $Na^+$  ○ $Al^{3+}$

**Table 5.1**
Composition of some important types of glasses

| Glass | Percentage of constituent oxides | | | | | | | Applications |
| | $SiO_2$ | $Na_2O$ | $K_2O$ | CaO | $Al_2O_3$ | $B_2O_3$ | Other | |
|---|---|---|---|---|---|---|---|---|
| Soda-lime | 72 | 13 | — | 10 | 1 | | MgO, 2 | Windows |
| Silica (fused) | 99.5+ | | | | | | | Thermal shock resistance |
| Lead silicate | 63 | 8 | 6 | 0.3 | 0.6 | 0.2 | PbO, 21 MgO, 0.2 | Electrical |
| High-lead | 35 | | 7 | | | | PbO, 58 | Crystal glass |
| Borosilicate | 80 | 4 | 0.4 | | 2.2 | 13 | | Chemical ware, hermetic seals |
| E-glass | 54 | 0.5 | | 22 | 14 | 8.5 | | Fiber reinforcement |
| Aluminosilicate | 57 | 1.0 | | 5.5 | 20.5 | 4 | MgO, 12 | High-temperature strength |

# 5.3   *Glass Compositions*

Commercial glasses are produced from inorganic oxides of which silica, or sand, is the most important constituent. These glasses are actually mixtures of oxides, not definite chemical compounds. The properties of the different constituents may be varied freely within certain limits, but if we exceed these limits, it may become difficult or impossible to form the glasses.

Compositions of some important glasses are given in Table 5.1, along with remarks about their special properties and applications. **Fused silica** glass is the most important single-component glass. It has high spectral and radar transmission and is not subject to radiation damage that causes browning of other glasses.

Soda-lime glass is the most commonly produced glass, accounting for about 90% of all glass produced. In this glass, the basic composition is 71–73% $SiO_2$, 12–14% $Na_2O$, and 10–12% $CaO$. The $Na_2O$ and $CaO$ decrease the softening point from 1600°C to about 730°C, which makes the soda-lime glass easier to form into shapes. To prevent crystallization, 1–4% $MgO$ is added to the glass and an addition of 0.5–1.5% $Al_2O_3$ is used to increase durability. Soda-lime glass is used as flat glass (windows, etc.), containers, and lighting products where high chemical durability and heat resistance are not needed.

---

## Case Study 5.1

### Fused Silica Aerospace Vehicle Windows

Fused silica is the only material that is suitable for aerospace missions as a combined optical and radar window. A search for other suitable compounds over a 30-year period has proved essentially fruitless.

Fused silica is currently used as the window in the space shuttle and as the radome and antenna window for several active missile vehicles. What makes this material so attractive is the combination of good optical transmission (with very low absorption when pure) and low coefficient of thermal expansion (which leads to low induced thermomechanical stresses under steep temperature gradients). Fused silica can withstand a temperature gradient of about 2000°C before reaching its fracture stress. Fused silica can also transmit radar in the microwave region at high temperatures with low dielectric loss. No other material can provide these useful characteristics as well as fused silica, which accounts for its extensive deployment in aerospace vehicles by aerospace engineers.

---

**Borosilicate** glasses are produced by the replacement of alkali oxides with boric oxide in the silica glass network. These glasses are characterized by lower coefficients of expansion. When $B_2O_3$ enters the silica network, it breaks the continuity of the silica-to-oxygen bonding and reduces the softening point of the silica glass. We attribute this effect to the presence of planar three-coordinate boron atoms in place of tetrahedral-coordinate silicon atoms. Borosilicate glass, because of its fairly low coefficient of thermal expansion ($4$–$5 \times 10^{-6}/°C$) is useful as heat-resistant laboratory containers, ovenware, sealed beam headlights, and so on.

Lead-based glasses are produced by replacing the lime of the soda-lime glass with lead oxide, PbO. Although the lime in soda-lime glasses must usually be limited to 15%, the lead oxide constituent in lead glasses can be increased to well over 60%. The lead oxide acts as a fluxing agent and tends to lower the softening point of the glass below that for the soda-lime type. It also improves the working qualities of the glass when the proportion of PbO is not more than 50%. Lead oxide–base glasses came into use some three centuries ago for decorative ware because of the brilliance that results from their high index of refraction. In such artistic forms, they are misnamed as *crystal* when indeed they are true glasses. Today, these glasses are still used for such aesthetic purposes as well as for shielding from high-energy radiation, for television tubes, fluorescent lamp envelopes, radiation windows, and even solder-sealing glasses.

Aluminosilicate glasses contain 20% or more alumina, smaller amounts of CaO or MgO, and very limited amounts of $Na_2O$ and $K_2O$. They are usually more difficult to melt and to work than the borosilicate glasses. They are characterized by high softening temperatures and relatively low coefficients of expansion that make them particularly suitable for higher temperature applications such as combustion tubes, higher temperature thermometers, and heating units.

Colored glasses are produced by adding various colorants, which are mainly transition metal oxides. The proportions are small, ranging from less than 1% to 3% or 4%, depending on the concentration of color desired and upon the tinting power of the colorant used. These colorants normally have little effect on the general physical properties of the base glass used. Solution colors are produced by dissolving metal oxides in glasses, making them capable of absorbing radiations of certain ranges of wavelength characteristic to the oxide used. Thus nickel produces a purple hue, cobalt a blue, chromium a green, uranium a greenish yellow, and ferrous iron a greenish blue. In addition, ferrous iron absorbs wavelengths in the infrared range so that such glasses can be used for absorption of heat radiation.

Colorless nonmetallic crystalline particles are used to produce translucence or almost **opaque** whiteness in glasses and glass coatings. These crystals, which have a different index of refraction from that of the base glass, scatter the light within the body of the glass and diffuse the transmitted light. Such **translucent** glasses are referred to as *opal* or *alabaster*. Most of these glasses depend on some constituent such as a fluoride or a phosphate that causes small particles to crystallize as the glass is cooled. Fluorides may be added in concentrations from 5% to 15%. Opal glasses are widely used for light diffusion in order to prevent glare. We use them for their decorative qualities in such items as kitchenware, dishes, and containers.

# 5.4   Viscosity of Glass

The **viscosity** characteristics of glasses are important from the practical standpoint of shaping products in glass manufacturing as well as in determining maximum permissible operating temperatures. The viscosity of glass increases continuously as it cools from the molten state to below the glass transition temperature, where small structural adjustments are still perceptible. The viscosity curves of several glasses listed in Table 5.1 are shown in Figure 5.4 together with viscosity reference points such as annealing point (to be discussed). The curves can be described by the following empirical expression:

$$\log_{10} n = -A + \frac{B}{T - T_g}$$

where $n$ is the viscosity at any temperature $T$, $T_g$ is the glass transition temperature, and $A$ and $B$ are constants determined from test data. We call this equation the Fulcher relation after its developer.

In the next section we will look more closely at glass forming. At this point, you should know that to shape any glass, it must flow, that is, the glass viscosity must be low by comparison to its rigid solid state. Specific viscosities in Figure 5.4 are for softening, which corresponds to a certain strain without fracture, annealing where cooling stresses can be dissipated by minor viscous atomic movement and the **strain point**, below which the glass is essentially rigid.

**Figure 5.4**
Temperature dependence of viscosity of glass (numbers refer to compositions given in Table 5.1)
(From W. F. Smith, *Principles of Materials Science and Engineering*, 2nd ed., McGraw-Hill Publishing Company, 1990.)

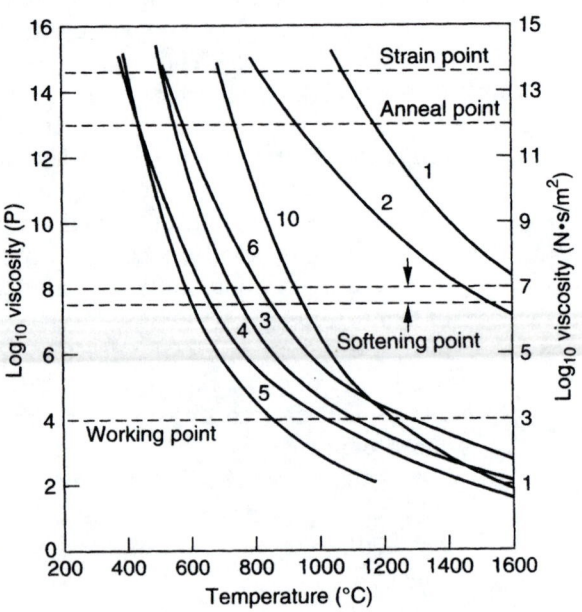

1 P (poise) = 1 dyn·s/cm$^2$;
1 Pa·s (pascal·second) = 1 N·s/m$^2$; 1 P = 0.1 Pa·s

For most commercial glasses the **annealing point** temperature is 35–40°C above that of the strain point. Glasses that have a large temperature interval between $10^5$ poises (see Figure 5.4) and the softening point are said to have a long **working range** and are generally easier to fabricate in the viscous state than glasses such as aluminosilicates that have a short working range. Although pure silica glass has a long working range as expressed in degrees, the working temperature is so high that the glass cools very rapidly and makes ordinary fabricating operations very difficult.

# 5.5 Glass Forming

Glass products are made by first heating the glass to a high temperature to produce a viscous fluid and then molding, drawing, or rolling it into a desired shape. We make sheet and plate glass by the float process, whereby a ribbon of glass moves out of a melting furnace and floats on the surface of a bath of molten tin, as shown in Figure 5.5. The glass ribbon is cooled while moving across the molten tin and while in a chemically controlled atmosphere. The molten tin produces an exceptionally smooth surface on the glass. When its surfaces are sufficiently hard, the glass sheet is removed from the furnace without being marked by rollers and is passed through a long annealing furnace in which residual stresses are removed.

Deep items such as bottles, jars, and light bulb envelopes are usually formed by **blow molding,** very much the same as for similar plastic items, as shown in Figure 5.6. Flat items such as optical and sealed beam lenses are made by pressing a plunger into a mold containing the molten glass. Funnel-shaped items such as television tubes are formed by centrifugal casting. Gobs of molten glass from the feeder are dropped onto a spinning mold that causes the glass to flow outward and upward to form a glass wall of the desired thickness.

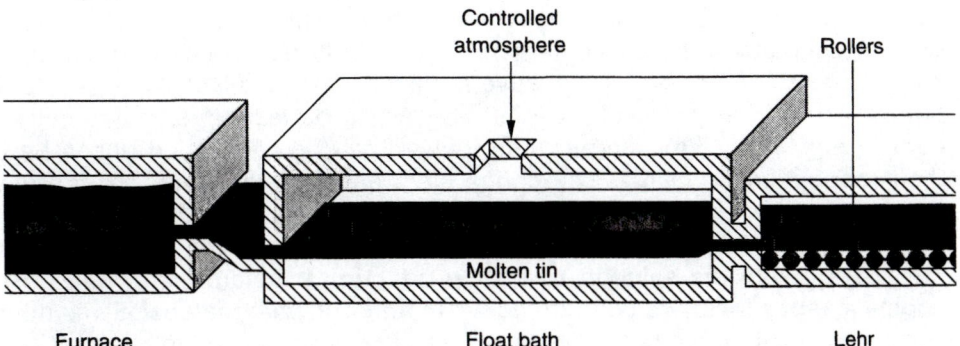

**Figure 5.5**
The float glass process
(From W. F. Smith, *Principles of Materials Science and Engineering,* 2nd ed., McGraw-Hill Publishing Company, 1990.)

**Figure 5.6**
Glass bottle blow molding

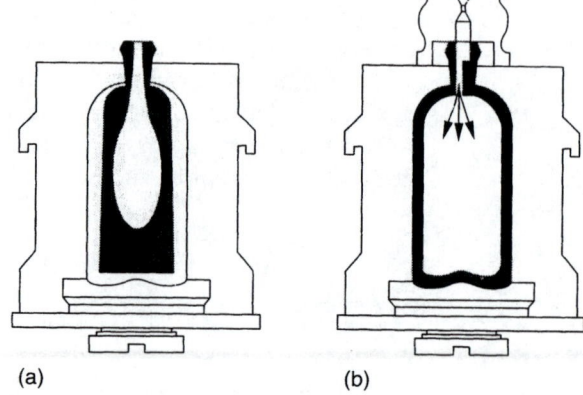

(a)                    (b)

## Case Study 5.2

### *Glass Provides an Environmentally Sound Solution to a Deburring Problem*

Vascol Industries produces machined and die cast components for many fabricators in the Northeast. Wet deburring is required to break all sharp edges and remove any undesirable raised metal. It has been general practice for many years to tumble the parts for one hour in a rotating drum containing small cones of polishing compounds embedded in a polymer. The practice has had many drawbacks, however, because the polymer cones wear and often get caught in the castings. Labor costs also have been high because frequent changes of the water-cone medium were required and environmental concerns necessitated treatment of the medium as an industrial waste because of the foaming of the polymer in the surfactants produced during tumbling.

Vascol managers decided to alter the deburring procedures. First, they evaluated electrostatic deburring but found that the polishing action that also occurred in this process was unacceptable for their customers. They finally approached the suppliers of the tumbling cones to determine if alternative materials were available. They evaluated four different cones, with unacceptable results. The last to be tested, left aside initially because the cone size was four times larger than that of the original, turned out to be the solution to their production problems. The cones were effective in deburring, but wore faster than other materials. The wear particles, however, were fine solid grit, which could be easily removed after draining the water, and no special disposal practice was required.

Happy with the solution to their production problems and their cost savings, one engineer sent a fractured cone to the lab to find out what material it was made from. To the surprise of all, it turned out to be a foamed silica glass, shown in Figure 5.7.

**Figure 5.7**
Foamed glass structure of deburring cone (300×)

The preparation of fiberglass is described in Chapter 7 on composites. This unique form of glass is produced by drawing molten glass through small-diameter orifices. Ultrapure silica fibers are the basis for telephone communication by optical light transmission.

**Tempered glass** is strengthened by rapid air cooling of the surface of the glass after it has been heated close to its softening point. The surface of the glass cools first and contracts, while the interior is warm and readjusts to the dimensional change with little stress. When the interior cools and contracts, the surfaces are rigid and so tensile stresses are created in the interior of the glass and compressive stresses on the surfaces. This tempering treatment increases the strength of the glass because applied tensile stresses must surpass the compressive stresses on the surface before fracture occurs. Tempered glass has a higher resistance to impact than annealed glass and is about four times stronger than annealed glass. Auto side windows and safety glass for doors are among items that are thermally tempered.

The strength of glass can also be increased by special chemical treatments. The replacement of sodium ions at a glass surface by larger potassium ions, accomplished by diffusion via a melted potassium salt in contact with the glass, can place the surface layer under compression with corresponding tensile stresses at the glass center. Since fracture most often originates at the surface, this technique is useful in strengthening assorted glassware products.

# Summary

Glasses are inorganic, rigid materials that are vitreous, or amorphous, that is, non-crystalline. Most glasses are based on a three-dimensional network of ionic and/or covalent bonded silica tetrahedra. Additions of other oxides, such as $Na_2O$, $K_2O$, $CaO$, and $MgO$, modify the silica network and reduce viscosity, allowing lower temperature processing. Important fabrication processes include flat plate manufacture by the float method and blow molding to produce container glass. Strength properties are improved by heat and/or chemical treatment to place exterior surfaces in compression. Glasses have very special properties such as transparency, hardness, good insulation characteristics, and high chemical resistance that make them important in many commercial products.

# Terms to Remember

| | |
|---|---|
| amorphous | opaque |
| annealing point | strain point |
| blow molding | tempered glass |
| borosilicate | transition temperature |
| brittle | translucent |
| fused silica | transparency |
| glass former | viscosity |
| glass modifier | vitreous |
| glass network | working range |
| melting | |

# Problems

1. Define glass and describe some of its properties.
2. What is a glass network former? Give two examples.
3. What is a glass modifier? Give two examples.
4. What is the most important constituent of commercial glasses?
5. Describe the properties of a pure fused silica glass.
6. What is the most commonly produced commercial glass?
7. What is the property of borosilicate glasses that makes them useful as heat-resistant laboratory containers?
8. Lead oxide–based glasses are often called *crystal*. How do you account for this?
9. How does glass viscosity vary with temperature?
10. How are sheet and plate glass fabricated?
11. What is tempered glass? Where is it used?
12. How can glass be strengthened after forming?

# 6

# *Ceramics Processing*

Artisans have been making ceramics since the beginning of mankind, using such naturally occurring powdered **raw materials** as clay, silica, and feldspar. Potter's wheels were devised millennia ago to shape the clay, which became moldable when water was added to it. The shapes that were created were dried by exposure then hardened by heating in a fire to produce containers mainly for water and food. (The word *ceramic* is derived from the Greek word *keramos*, which translates as *burnt stuff*.) Clay fuses, or melts, over a range of temperatures because the amount of liquid phase gradually increases with increasing temperature. Therefore, a dense strong ceramic piece is produced during firing without complete melting so that the desired shape, with some **shrinkage**, is maintained.

## 6.1 Traditional Ceramics

In addition to clay, ceramic artisans used nonclay minerals such as silica (sand), finely ground flint or quartz, and a flux such as feldspar. These raw materials used by ancient civilizations are still used today in traditional ceramic manufacturing. Many of the products of this industry are outlined in Table 6.1

We cannot overemphasize the importance of **clay** in the evolution of ceramic processing. The **plasticity** that clay develops when water is added to it provides the **workability** necessary to make pottery, dinnerware, brick, wall tile, and piping at low cost. We use silica as a filler because it is inexpensive, hard, and chemically inert at temperatures needed for **densification,** which eliminates porosity. Silica experiences little change during high-temperature processing because it has a high **melting temperature** (when it does melt, it forms silica glass with very high viscosity).

When mixed with clays, feldspar acts as a **flux** to form glasses that have relatively low melting temperatures. Feldspar minerals are aluminosilicates that range in **composition** from $KAlSi_3O_8$ to $NaAlSi_3O_8$ to $CaAl_2Si_2O_8$. We use them to reduce the densification temperatures of traditional ceramics by supplying a low-temperature liquid phase.

In traditional ceramics, mined raw materials usually go through a milling or grinding operation to break up agglomerates or reduce **particle size**. This is followed by screening or sizing to provide a powder with the desired range of particle sizes. **Mixing** distributes different materials uniformly to prepare the mixture for the next processing step. For complex combinations, powders are thoroughly mixed, usually with water and additives to impart flow characteristics that are compatible with forming methods.

Of course, the properties and characteristics of traditional ceramic products are affected by their composition, but in industry, we must deal with large amounts of each component—clay, silica (quartz), and feldspar. We use three-dimensional ternary **phase diagrams** to describe these complicated compositions and phases but, fortunately, we simplify them by taking triangular slices at constant temperature. Figure 6.1 illustrates such a slice in the clay (mullite)-silica-leucite (flux) system that is used for many commercial porcelains. In this figure, the dark lines represent the eutectic temperature changes as ternary **additions** are made, and the lighter lines represent constant temperatures along the surface that define the melting temperature. Shaded areas represent typical compositions of commercial whiteware.

**Table 6.1**
Traditional Ceramics

| Product | Application |
| --- | --- |
| Whiteware | Dishes, plumbing, enamels, tiles |
| Heavy clay products | Sewer pipe, brick, pottery, sewage treatment, and water purification components |
| Refractories | Brick, castables, cements, crucibles, molds |
| Construction | Brick, block, plaster, concrete, tile, glass, fiberglass |
| Abrasive products | Grinding wheels, abrasives, milling media, sandblast nozzles, sandpaper |

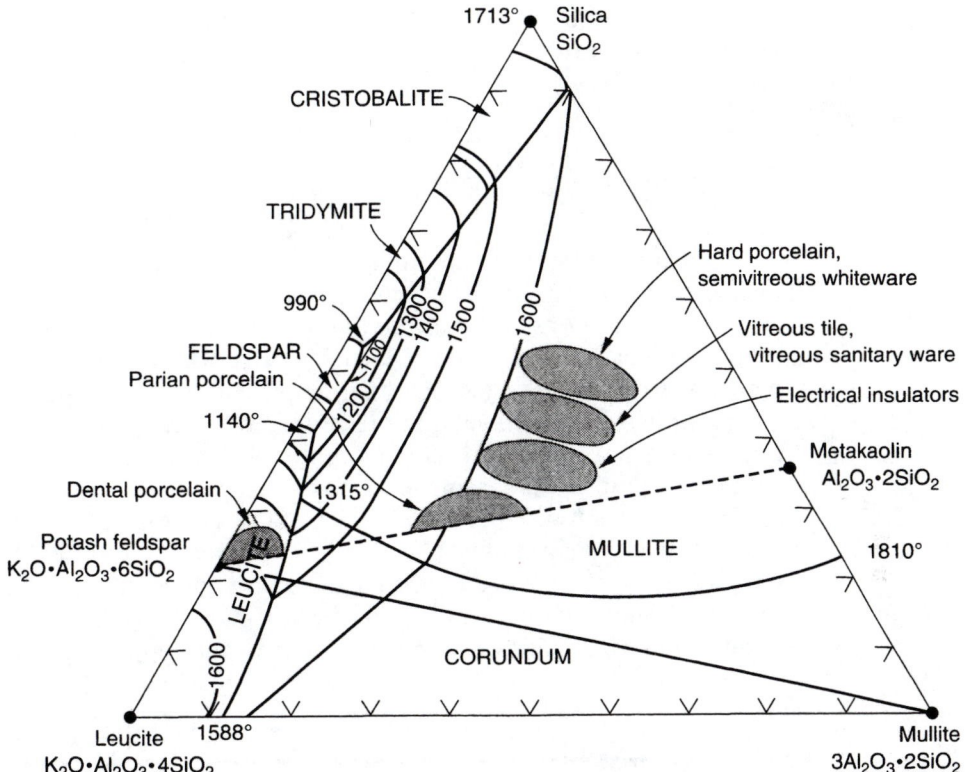

**Figure 6.1**
Areas of whiteware compositions shown on the silica-leucite-mullite phase-equilibrium diagram
(From W. D. Kingery, H. K. Bowen, and D. R. Uhlmann, *Introduction to Ceramics,* 2nd ed., John Wiley & Sons, 1976.)

If we consider the dashed line in Figure 6.1, we can visualize the effects of composition on processing much better. Along this line, which is really the liquidus projection of a binary "alloy," the addition of feldspar to metakaolin clay lowers the liquidus until it reaches the ternary eutectic (dark line from 1315°C to 1810°C). This means we can densify, or fire, the composition at lower temperatures with increasing amounts of flux. On the other hand, if the composition has greater clay content, it has better molding characteristics but must be fired at higher temperatures. For example, dental porcelains are highly translucent and are made in small, relatively simple shapes, so we use high-feldspar/low-clay mixtures. By contrast, porcelain art-ware and tableware frequently have complex shapes, and the ability to form them is more important than the firing temperature. They can be formed by hand throwing on a potter's wheel, by slip casting using a slurry made by mixing in large amounts of water, or by jiggering, which is a mechanized form of the potters wheel whereby the clay is rotated while being pressed against the wall of a shaped mold by a profiled tool. We use compositions with high clay content for these applications.

# 6.2 Modern Ceramics

We have learned a lot from traditional ceramics and now have a much better understanding of the interaction between ceramic compounds and their processing. The synthesis of high-purity materials and an understanding of the role that impurities play in densification and final properties has led to a generation of new ceramics referred to as modern, or technical, ceramics. With some exceptions, they typically have controlled compositions that are prepared by chemical synthesis rather than being made from naturally occurring minerals. These materials fill the needs of applications that are too demanding for traditional ceramics. Modern ceramics include the oxides $Al_2O_3$, $MgO$, $ZrO_2$, $BeO$, $SiO_2$, and $MgAl_2O_4$ for electrical, structural, and optical applications; $UO_2$ and $UC$ for nuclear fuels; $BaTiO_3$ for electronic usage; ferrites such as $ZnFe_2O_4$ for magnetic, low-loss ceramics; and the refractory carbides, nitrides, and borides for advanced high-temperature structural materials and tools. Table 6.2 summarizes many modern ceramic applications.

**Table 6.2**
Modern Ceramics

| Industry | Application | Composition |
|---|---|---|
| Electric | Magnets, insulators, capacitors, transducers, dielectrics, lasers, substrates, semiconductors, sensors, heating elements | $Al_2O_3$, $MgO$, $BeO$, $BaTiO_3$, $SiO_2$, $ZnFe_2O_4$, $ZnO_2$ |
| Aerospace | Heat shields, IR domes, radomes, rocket motor nozzles | $SiO_2$, $Si_3N_4$, $Al_2O_3$, $MgAl_2O_4$, carbon fiber components |
| Automotive | Catalytic converters, oxygen sensors, advanced engines, turbocharger rotors | Low-expansion silicates, $Si_3N_4$, $SiC$, $ZnO_2$, $Al_2O_3$ |
| Armor | Light-weight small arms armor, transparent and opaque | $B_4C$, $SiC$, $Al_2O_3$, $MgAl_2O_4$, single-crystal $Al_2O_3$ |
| Medical | Bone, joint, and tooth replacement | $Al_2O_3$, silicates |
| Nuclear | Fuel and control rods, waste containment and solidification | $UO_2$, $UC$, $B_4C$, various glass formulations |
| Metal processing | Cutting tools, dies, and molding materials | $TiC$, $WC$, $ZnO_2$, $Al_2O_3$, $TiN$, $SiC$, $Si_3N_4$ |

# 6.3 Raw Material Processing

The most critical factors affecting forming and firing processes are the raw materials and their preparation. Particle size and particle size distribution play very important roles in achieving desired forming characteristics and final fired density. Typical clay materials have a particle size distribution that ranges from 0.1 $\mu$m to 50 $\mu$m for the individual particles. For the preparation of porcelain compositions, the flint and feldspar constituents have a substantially larger particle size ranging between 10 $\mu$m and 200 $\mu$m. The fine-particle constituents of modern ceramics are often less than 1 $\mu$m to facilitate cold-forming processes because these depend on small particles flowing over one another or remaining in a stable suspension.

Impurity control in raw materials plays a major part in the successful development of ceramic products, particularly modern ceramics, as pointed out in Case Study 6.1.

---

## Case Study 6.1

### Ceramic Aluminum Oxide Gas Bearings for Aerospace Vehicles

In the late 1960s, certain compounds such as $Al_2O_3$, BeO, and $B_4C$, which are low-density ceramics with a high elastic modulus and excellent wear resistance characteristics, were potential replacements for metal gas bearings in gyroscopes. The higher modulus-to-density ratios of the ceramics offered improved operational characteristics over the metals, but there was a requirement that the surfaces of the ceramics be highly polished and have absolutely no impurity particles on exposed contact surfaces. This restriction meant that the material needed to have very fine grain size, no porosity, and no impurity inclusions. At that time, the best choice to meet the requirements was $Al_2O_3$.

TV Fabricators, a large bearing manufacturer, obtained the best fine-particle $Al_2O_3$ powder available (about 99.95% pure) and proceeded to fabricate billets from which gyroscope components could be machined. All the requirements were met except for inclusions that showed up intermittently on exposed surfaces. Analysis revealed that these were silica, $SiO_2$, particles that were present in the aluminum oxide powder as supplied by the raw material producer. Although the overall purity was 99.95%, there were still enough particulate silica particles present that many completed gyroscope parts were unacceptable. The powder supplier offered no corrective measures since this market was small and "other users of the powder were satisfied with it." TV Fabricators had to solve the problem on their own.

After several failed attempts, they finally succeeded by using an air classifier to separate the coarser silica particles from the finer sized $Al_2O_3$. A schematic of such a facility is shown in Figure 6.2. They separated coarse and fine dry powders by controlling horizontal centrifugal force and vertical air currents within the classifier. Particles enter the equipment along the centerline and are centrifugally accelerated outward. As the coarse particles move radially away from the center into the separating zone, they slow down and settle into a collection cone. The finer particles are carried upward and radially by the air currents

**Figure 6.2**
Drawing of an air classifier,
showing the paths of the
coarse and fine particles
(Courtesy Sturtevant, Inc.,
Boston, Mass.)

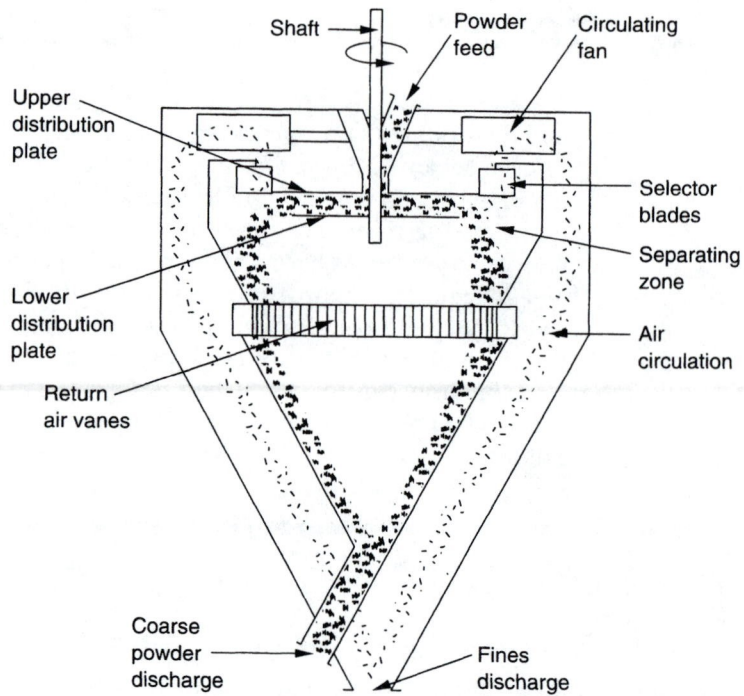

through selector blades to a separate cone for collection. For bearing material, separation was carried out at about 20 $\mu$m. Everything larger than this size (including silica contamination) was discarded, and the fraction smaller than 20 $\mu$m (most of the pure alumina) was used to successfully fabricate gas-bearing-quality $Al_2O_3$ with no reject part.

Technology changes, however, and today modifications in chemical processing have resulted in production of several grades of very pure aluminum oxide without the need for such follow-on procedures.

**Air classification** is frequently linked directly to milling, crushing, grinding, and other comminution equipment in a closed circuit. Particles from the mill are discharged directly into the air classifier. The fine particles are separated and the coarse fraction is returned to the mill for further grinding.

Ceramic powders prepared for the modern ceramic industry are synthesized by batch process operations. Powder specifications for batches are often available to users, but selection is usually based on a trial basis; for example, large quantities of $Al_2O_3$ are used in the manufacture of high-temperature sodium vapor lamp envelopes. Figure 6.3 presents a photograph of such a lamp and a corresponding photomicrograph of the dense **microstructure**. Although a material specification must be met by powder suppliers, acceptance of new powder lots is more often based on results of the complete process fabrication and densification of a sample lot. This is done because everything that may have an important bearing on the quality of the final product is not specified for original powder. The test-sample process run serves to avoid unpleasant surprises. In general, ceramic materials powder specifications include particle size and distribution, chemical impurity levels, and crystal phases that are present.

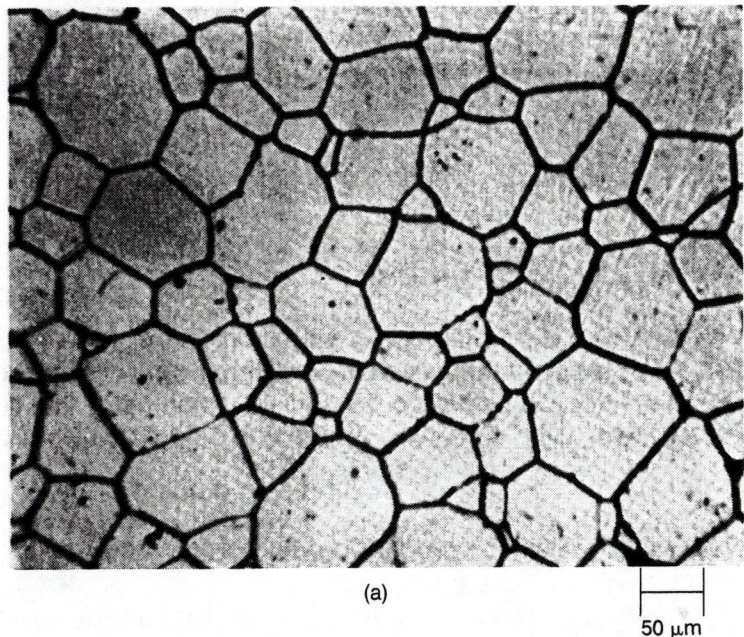

(a)

50 μm

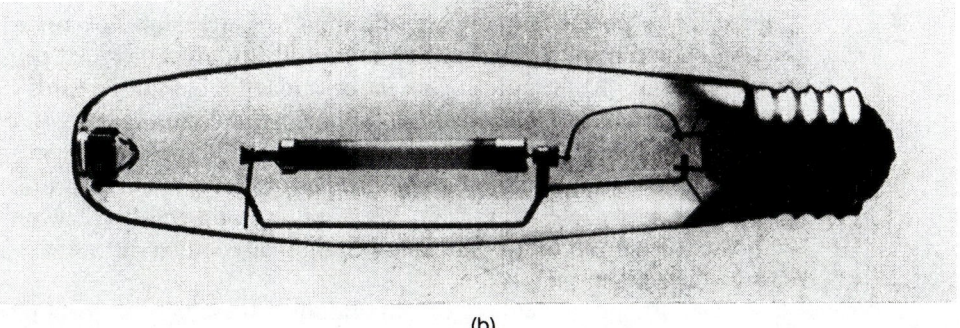

(b)

**Figure 6.3**
Translucent Lucalox® alumina: (a) pore-free microstructure necessary for translucence, (b) sodium vapor lamp used in energy-efficient sodium streetlights
(From J. S. Shackelford, *Introduction to Materials Science for Engineers,* 3rd ed., Macmillan Publishing Company, 1992.)

Typical chemical reactions that have been employed to produce high-quality powders for modern ceramics fabrication include the following:

$$2\ Al(OH)_3 \rightarrow Al_2O_3 + 3\ H_2O$$
$$MgCO_3 \rightarrow MgO + CO_2$$
$$MgO + Al_2O_3 \rightarrow MgAl_2O_4\ (spinel)$$
$$SiCl_4 + CH_4 \rightarrow SiC + 4\ HCl$$
$$3\ Si + 2\ N_2 \rightarrow Si_3N_4$$
$$4\ B + C \rightarrow B_4C$$
$$BaCO_3 + TiO_2 \rightarrow BaTiO_3 + CO_2$$

# 6.4  Fabrication Methods

We make useful shapes from polycrystalline technical ceramic powders by consolidating them, followed by firing, which densifies or sinters them into a strong solid. Three fabrication processes that have been used for many years include casting, extrusion, and dry pressing. We often distinguish these processes by the volume of water that each requires for shape forming.

## 6.4.1  Casting Processes

**Casting** is a familiar process that employs a stable suspension of particles in a fluid, usually water in the range 25–50 volume percent. The suspension, called a *slip*, is poured into a porous mold where the fluid from the slip is absorbed into the mold leaving behind a solid layer on the mold wall (the thickness, of course, depends on length of time in the mold). This process may be continued until the entire mold cavity becomes solid (solid casting), as demonstrated in Figure 6.4, or we may end it when the wall thickness reaches a specified dimension, pouring out any excess slip. This last procedure is called *drain casting* (also shown in Figure 6.4). As the cast piece dries, it shrinks away from the mold wall and can be removed by tapping to extricate it from the mold. Molds are often made in two or more sections to ensure easy cast specimen removal.

The primary advantage of slip casting is the ability to form intricate shapes at relatively low cost. Figure 6.5 shows a photograph of experimental turbine engine rotors of silicon nitride fabricated by this process. Most molds are made from pottery plaster using the approximate ratio of three parts water to four parts plaster, by weight. The plaster of paris ($CaSO_4 \cdot {}^1/_2 H_2O$) rehydrates to form a network of needle-shaped gypsum crystals that are separated by pores a few tenths of a micron in diameter. These small pores create a suction that draws liquid from the slip.

**Figure 6.4**
The steps in using a plaster of paris mold: (a) solid slip casting, (b) drain slip casting (From W. D. Kingery, *Introduction to Ceramics*, John Wiley & Sons, 1960.)

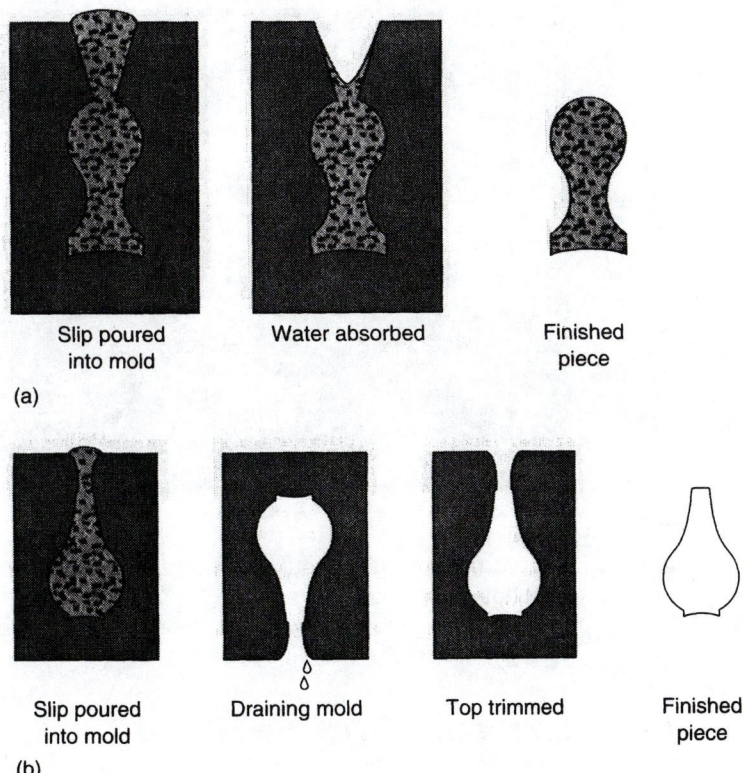

Slip poured into mold

Water absorbed

Finished piece

(a)

Slip poured into mold

Draining mold

Top trimmed

Finished piece

(b)

Characteristics of a good casting slip include the following:

- *Viscosity*   The slip must pour easily and fill all details of a mold, allowing air bubbles to rise and break.
- *Settling rate*   Particles in suspension must not settle appreciably during the time necessary to make a casting (otherwise tapered walls and thick bottoms will result).
- *Release*   The cast piece should shrink slightly during drying to allow release, and the dry cast must release from the mold at the proper time (premature release causes an uneven wall thickness). The lack of proper release during drying can cause stresses in the wall or may result in fracture during removal.
- *Strength of cast ware*   The cast object must have sufficient strength to allow handling, trimming, or further shaping operations.

**Figure 6.5**
Si₃N₄ pressure cast gas turbine rotors before (light) and after (dark) glass encapsulation and hot isostatic pressing
(From D. W. Richerson, *Modern Ceramic Engineering: Properties, Processing, and Use in Design,* Marcel Dekker, Inc., 1992.)

# 6.4.2 Doctor Blade Process

The **doctor blade** process is a form of casting whereby divided powders are suspended in aqueous or nonaqueous liquids that contain solvents, plasticizers, and binders to form a slurry that is cast onto a moving carrier surface. The slurry passes beneath the knife edge of a blade that levels, or "doctors," it into a layer of controlled thickness and width as the carrier advances along a supporting table. When the solvents evaporate, fine solid particles coalesce into a relatively dense, flexible sheet that may be stored on take-up reels or stripped from the carrier in a continuous sequence. This process is shown for tape casting in Figure 6.6. Tape casting is a continuous, high-productivity process that produces thin, flat ceramic substrates with smooth surfaces. Substrates for multilayer ceramic electronic packaging, multilayer barium titanate capacitors, piezoelectric devices, thick and thin film insulators, ferrite memories, and catalyst supports are made this way.

**Figure 6.6**
Doctor blade tape casting
(From D. W. Richerson, *Modern Ceramic Engineering: Properties, Processing, and Use in Design,* Marcel Dekker, Inc., 1992.)

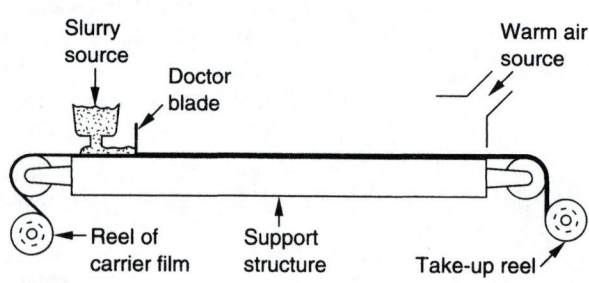

## 6.4.3 Dry Pressing

The most popular method for fabricating technical ceramic components as well as more traditional structural refractories is **dry pressing**. This process involves the simultaneous compaction and shaping of a granular powder containing small amounts of water and/or organic binders. The inherent value of dry pressing comes from its ability to rapidly form a wide variety of shapes with close tolerances and the potential for automation. Products made by pressing include magnetic and dielectric ceramics, various fine-grained technical ceramics such as spark plugs, cutting tools, refractory oxygen sensors, certain ceramic tile and porcelain products, coarse-grained refractories, grinding wheels, and structural clay products.

Stages in dry pressing, shown in Figure 6.7, include (1) filling the die, (2) compaction and shaping, and (3) ejection. Free-flowing granules fed to the die by means of a sliding feed shoe are metered volumetrically. Feeding typically is synchronized with a drop in the bottom piston followed by a compression step and then ejection of the piece. The die materials vary from hardened steels for relatively soft powders to metal carbides for very abrasive powders. Clearance between die and punch is important, is dependent on powder size, and the die wall is sometimes tapered to facilitate ejection. Pressing times vary from a fraction of a second for small parts to several minutes for large pieces, using a single-action press. Rates exceeding 5000 parts per minute are achieved using a multistation rotary press. Press capacities range up to hundreds of tons and the pressing pressure used is commonly in the range of 20–100 MPa. Higher pressures are employed for technical or modern ceramics than for traditional clay-based composites.

Ceramic powders are granulated with such additions as binders, plasticizers, and lubricants, usually by a spray-drying method to ensure reproducible volumetric filling and uniform density of filling as well as to provide strength to the compact with minimal density gradients. The use of solid lubricants such as stearic acid also minimizes die wear. Figure 6.8 shows micrographs of industrial ceramic pressing granules with spherical shapes and with a wax lubricant mixed in.

Pressure produced by the moving punches compacts the granules into a cohesive shape. Sliding and rearrangement of particles within granules reduces porosity and increases the number of intergranular contacts. Air compressed in pores migrates and partially exhausts between the punch and die. A typical fill density is in the range of 25% to 35% of theoretical density, and the compaction ratio, that is, the ratio of pressed density to fill density, is less than two, corresponding to compacted densities not much higher than about 50%. This is in contrast to ductile metal particles such as aluminum and copper that can be cold pressed to almost 100% of theoretical density at pressures of about 500 MPa.

## 6.4.4 Isostatic Pressing

Pressed products that have one elongated dimension, a complex shape, or a large volume are not easily pressed in dies but are often produced by the technique of

**Figure 6.7**
Schematic representation of the steps in uniaxial powder pressing: (a) the die cavity is filled with powder, (b) the powder is compacted by means of pressure applied to the top die, (c) the compacted piece is ejected by rising action of the bottom punch, (d) the fill shoe pushes away the compacted piece and the fill step is repeated
(From W. D. Kingery, editor, *Ceramic Fabrication Processes,* MIT Press, 1958.)

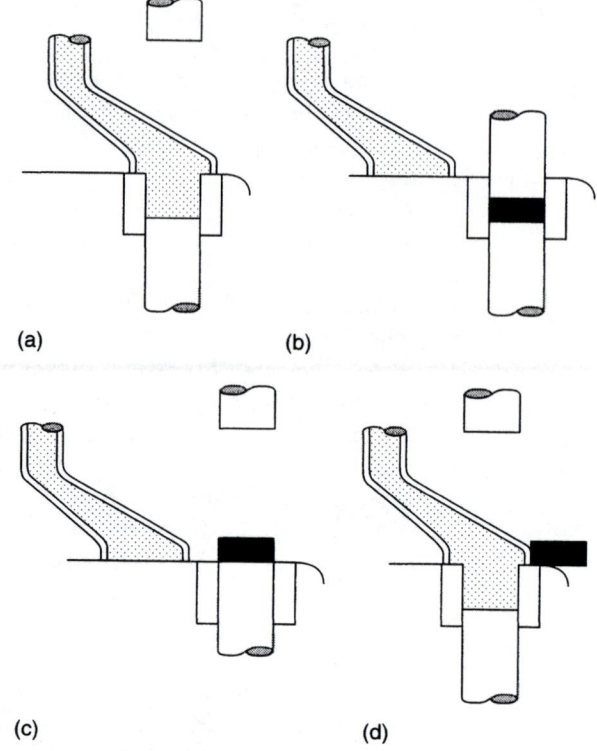

(a)                    (b)

(c)                    (d)

**isostatic pressing**, whereby pressure is applied from all sides instead of unidirectionally. If a rubber mold is filled with dry powder then inserted into liquid, hydraulic pressure on the filled mold is equal in all directions. Normal wall friction is eliminated, resulting in more uniform density of the compacted material. During firing to achieve final densification, shrinkage is more uniform with much less tendency toward warping and cracking. Steps in isostatic pressing are shown in Figure 6.9 for a wet-bag process, whereby flexible rubber molds are filled, sealed, and then pressed isostatically. When pressure is released, the molds are removed and parts are ejected. Any noncompressible fluid can be used for isostatic pressing, but water with a rust inhibitor is the most common.

Laboratory isostatic presses have been built with pressure capabilities ranging from 35 to 1380 MPa; however, production units usually operate at 400 MPa or less. Although wet-bag pressing can provide uniform density in a wide variety of types and sizes of parts, the disadvantages are that it is a batch process with a long cycle time, high labor requirement, and low production rates.

Dry-bag isostatic pressing was developed to achieve increased production rates with improved control of dimensional tolerances. Rather than immerse the tooling (the assembly of jigs, fixtures, etc.) in a fluid, we build tooling with internal channels into which we pump high-pressure fluid. With a properly designed system, parts can be pressed at a rate of 1000 to 1500 cycles per hour. Dry-bag pressing has been used

**Figure 6.8**
Micrographs of (a) industrial ceramic pressing granules with spherical shapes and (b) pressing granules with a wax lubricant mixed in (From J. S. Reed, *Introduction to the Principles of Ceramic Processing,* John Wiley & Sons, 1988.)

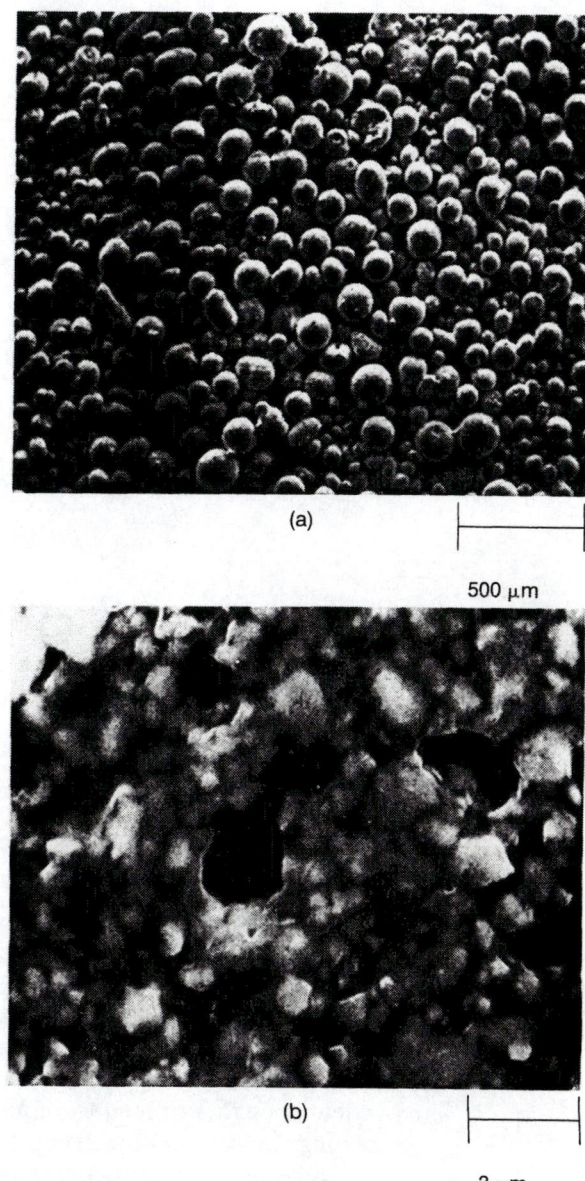

(a)

500 μm

(b)

3 μm

for many years to press spark plug insulators, as shown in Figure 6.10. Multiple sections are built into a single mold and the pressurized fluid enters carefully positioned channels in the elastomer envelope. Isostatic pressing is used to fabricate parts with compound curvature as well as large configurations that cannot easily be pressed uniaxially in dies. Figure 6.11 shows a variety of parts fabricated by uniaxial and isostatic pressing. The larger parts in the upper part of this figure were made by isostatic pressing.

**Figure 6.9**
Schematic of a wet-bag isostatic
pressing system
(From D. W. Richerson, *Modern
Ceramic Engineering: Properties,
Processing, and Use in Design,*
Marcel Dekker, Inc., 1992.)

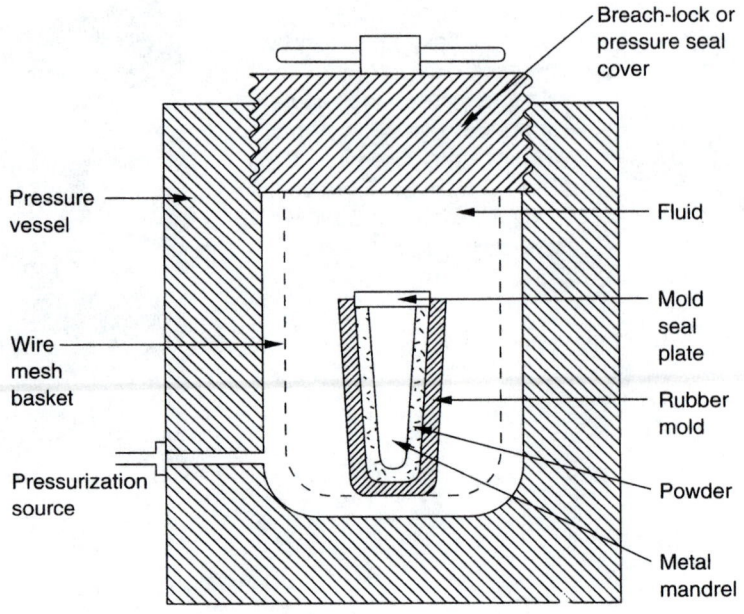

## 6.4.5  Plastic Forming

Many traditional ceramic compositions based on clay are formed by jiggering, plastic pressing, extrusion, and injection molding. These processes often require nothing more than water in the range of 15–30 volume percent to provide adequate plastic behavior. Modern ceramic compositions also are formed into shapes by **plastic forming**, using organic binders and plasticizers as well as water. Often as much as 25 to 50 volume percent organic additive is required. Jiggering, remember, involves a rotating mold to form one surface of the product; it is used to prepare items such as tableware and pottery that have axes of circular symmetry. Other methods of plastic forming include hand pressing and sculpting by artists working with ceramics.

In general, plastic forming requires a fine-particle paste with sufficient yield strength so that the formed product will maintain its shape during handling. Excessive yield strength can lead to rupture and yield strength is effectively controlled by regulating the amount of water and/or plasticizer that is used.

**Extrusion** is a plastic-forming method we use to fabricate long-length ceramic parts with a constant cross-sectional area. Such items as bricks, pipes, circular insulators, and capacitors are extruded. The process consists of using an extrusion machine, such as that shown in Figure 6.12, to force the plastic ceramic mixture through a shaped die.

Material to be processed is fed to the pug mill chamber where paddles or augers mix and shred it into a vacuum chamber. The de-aired material is consolidated and extruded continuously using the auger and extrusion die. Complex dies

**Figure 6.10**
Pressing mold used in dry-bag pressing
(From D. W. Richerson, *Modern Ceramic
Engineering: Properties, Processing, and
Use in Design,* Marcel Dekker, Inc., 1992.)

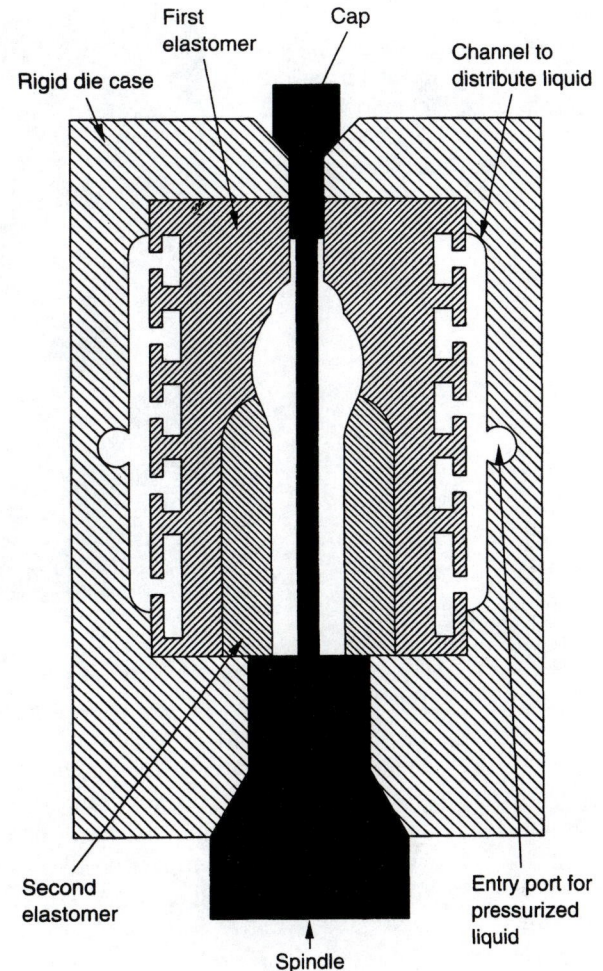

may contain small channels for injecting a die wall lubricant. For the conventional extrusion of hollow items, a system of arms, called a *spider*, supporting the central core rod is attached to the die and the arms interrupt the flow. Flow through the finishing tube improves the knitting of material separated by the spider. The extrusion die must generate an internal pressure and flow pattern that eliminates cavities, defects, and so on. Industrial extrusion pressures range up to about 5 MPa for clay water-base compositions and up to 15 MPa for some organically plasticized materials. Capacities range widely, depending on product size, but may approach 100 tons per hour for large products. Industrial extrusion rates of one meter per minute are common for many clay-base porcelain and whiteware compositions. Examples of some extruded shapes are shown in Figure 6.13.

**Figure 6.11**
Variety of parts fabricated by uniaxial and isostatic pressing (From D. W. Richerson, *Modern Ceramic Engineering: Properties, Processing, and Use in Design,* Marcel Dekker, Inc., 1992.)

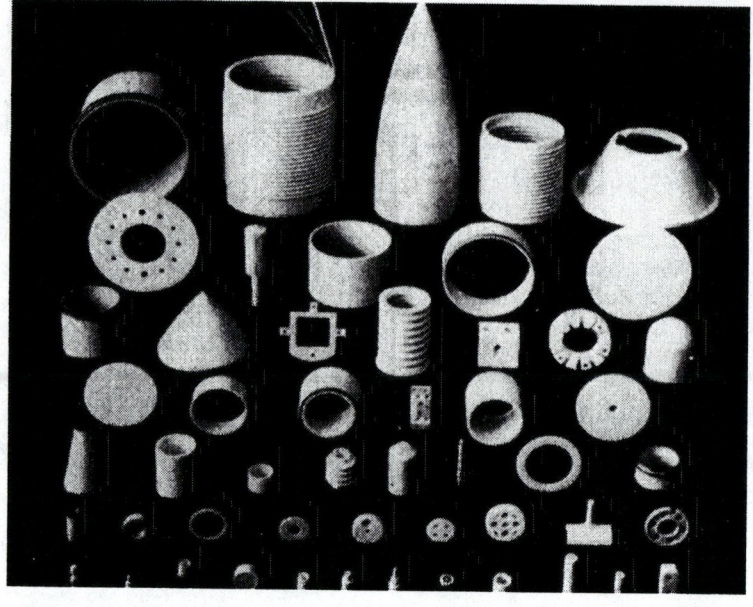

The **injection molding** of ceramics is similar to injection molding of polymers. Feed material consists of a mixture of ceramic powder and a thermoplastic polymer containing a plasticizer, wetting agent, and antifoaming agent. The mixture is preheated in the barrel of the injection-molding machine to a temperature suitable for pressurized flow, then the plunger is pressed against the heated material, forcing it through an orifice into a shaped die cavity. The mixture is fluid at this point with no yield strength (in comparison to an extrusion mix that has enough yield strength to form self-supporting shapes). A part can be removed from its die as soon as it is

**Figure 6.12**
An auger-type ceramic extruder (From D. W. Richerson, *Modern Ceramic Engineering: Properties, Processing, and Use in Design,* Marcel Dekker, Inc., 1992.)

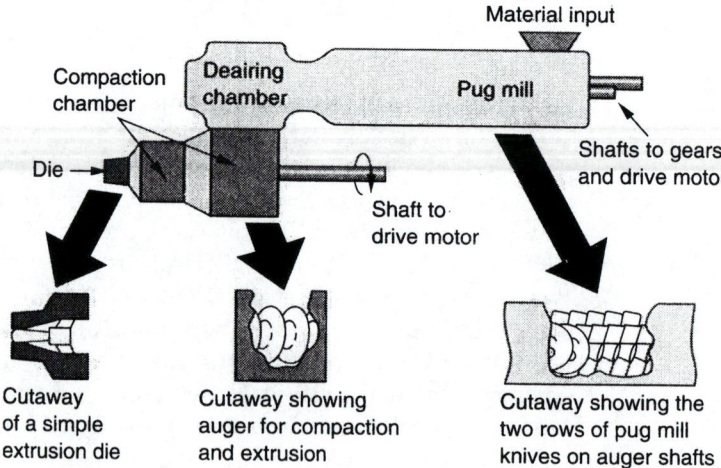

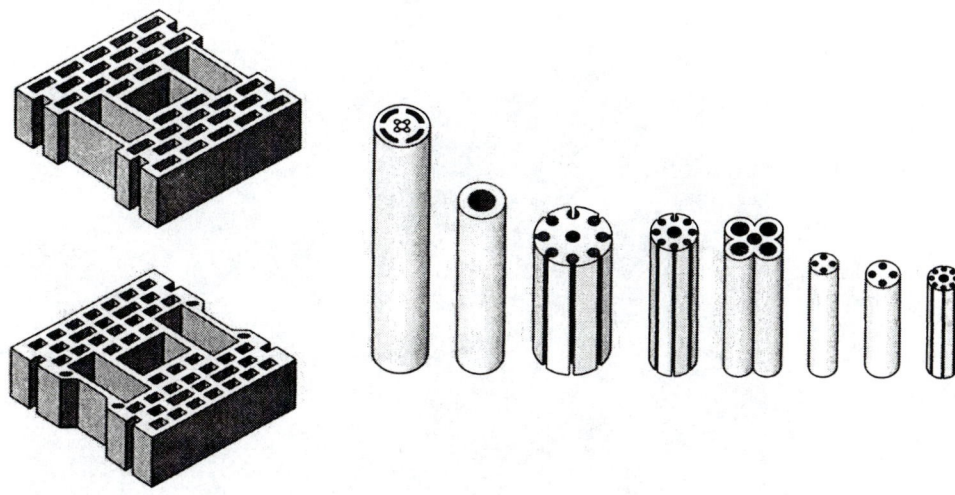

**Figure 6.13**
Examples of the types of shapes that are commonly extruded
(From D. W. Richerson, *Modern Ceramic Engineering: Properties, Processing, and Use in Design,* Marcel Dekker, Inc., 1992.)

rigid enough to handle without causing distortion. Cycle times can be rapid, providing the potential for high-volume, low-cost capability to make ceramics with complex shapes. Ceramic parts are made with the same injection-molding equipment as used by the plastics industry but with dies that are made from more wear-resistant metal alloys. The thermoplastic filler in the ceramic powder batch occupies up to 40 volume percent of the mixture and after shape forming, the plastic is removed by slow, careful heat treatments, making it a major factor in the successful fabrication of parts, because the large volume of organic material must be removed without cracking or distorting the ceramic. Examples of various ceramic parts formed by injection molding are shown in Figure 6.14.

# 6.5   *Densification by Sintering* _____

Densification, the final step in the fabrication of ceramic parts, is accomplished by **sintering,** a process that involves atom diffusion, viscous flow, evaporation, and condensation mechanisms. We can think of it, however, simply as the removal of porosity that is accompanied by shrinkage to form a dense structure. Stages that the material goes through during sintering are illustrated schematically in Figure 6.15. This figure shows spherical particles for convenience, but the basic mechanism is very much the same for different particle sizes and for irregularly shaped particles.

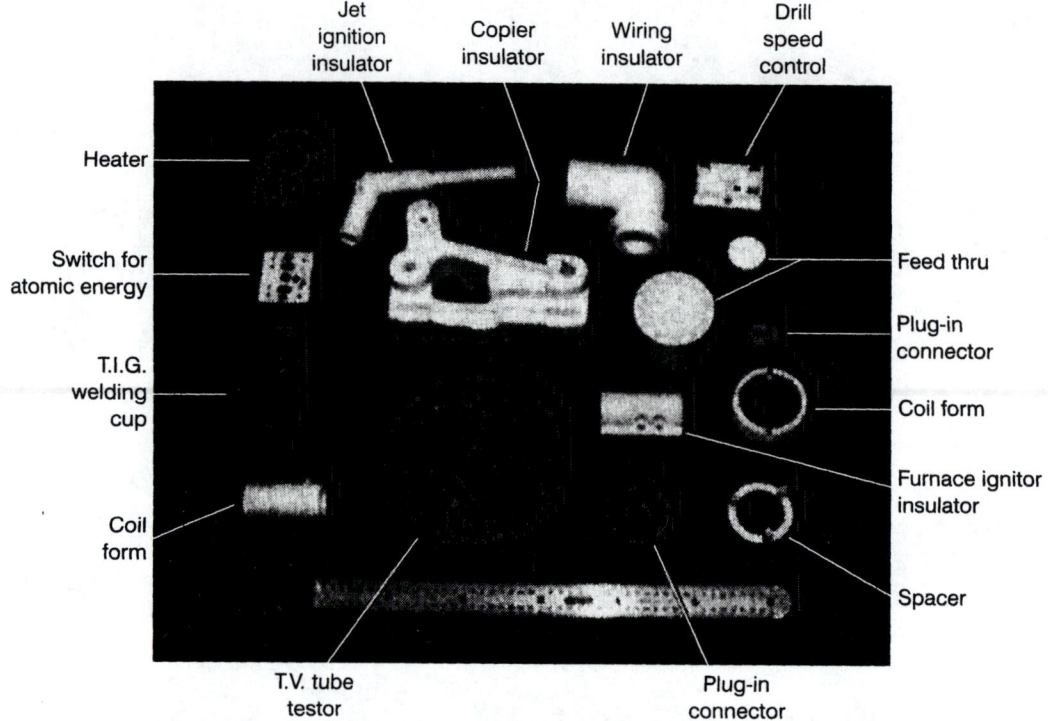

**Figure 6.14**
Examples of $Al_2O_3$ parts fabricated by injection molding
(Courtesy D. W. Richerson, *Modern Ceramic Engineering: Properties, Processing, and Use in Design,* Marcel Dekker, Inc., 1992.)

After cold forming, the powder particles are in surface contact with one another. During the initial stage of sintering, neck growth occurs along the contact region between adjacent particles with the simultaneous development of particle boundaries (now called grain boundaries) between particles and pores in the interstices. As sintering progresses, the pores become smaller, usually more spherical in shape, the contact region grows, and overall porosity decreases as the centers of the original particles move closer together. This process results in shrinkage equivalent to the amount of porosity reduction.

The influence of these parameters on densification has been investigated more thoroughly for modern ceramic materials, where control of chemical composition, purity, and particulate characteristics are vital to densification and final properties. In recent years, for example, translucent and transparent ceramic oxides, such as $MgO$, $Al_2O_3$, $MgAl_2O_4$, $BaTiO_3$, and $Y_2O_3$, have been made possible by solid-state sintering using careful control of fine particle sized starting powders (usually of submicron dimensions) coupled with control of the grain growth during sintering by using appropriate additives.

**Figure 6.15**
Structural changes during sintering: (a) pressed spherical powder, (b) coalescence and pore formation, (c) densification

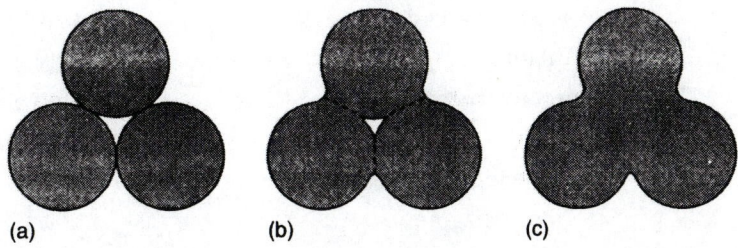

(a)  (b)  (c)

The process of liquid-phase sintering is more often employed for traditional ceramics. During firing, a low-melting-point phase melts, forming a viscous siliceous liquid that acts both as an adhesive for bonding and as a fluid that flows into pores, effectively sintering the solid particles. For satisfactory firing, we must control the amount and viscosity of the liquid phase present. Particle size, time, and temperature are important in this case as well as for technical or modern ceramics that do not involve a liquid phase. Shrinkage generally occurs with sintering and can range from as low as 10% to as much as 50% by volume and is a function of initial preform density.

# Summary

The processing of crystalline ceramic materials involves the agglomeration of small particles in dry, plastic, or fluid states by a variety of methods. Cold-forming processes such as dry pressing, extrusion, and slip casting are common methods to produce preforms or "greenware."

After cold forming, ceramics are densified by sintering at elevated temperatures to produce a solid, with a major reduction in porosity. In certain cases, pore-free ceramics have been prepared by careful control of such process parameters as chemical composition and firing or sintering conditions. Whereas technical ceramics are most often sintered by solid-state diffusional processes, traditional ceramic products often employ the presence of a liquid phase to facilitate densification by capillary action.

# Terms to Remember

| | |
|---|---|
| additions | doctor blade |
| air classification | dry pressing |
| casting processes | extrusion |
| clay | fabrication |
| composition | flux |
| densification | injection molding |

isostatic pressing
melting temperature
microstructure
mixing
particle size
phase diagram

plastic forming
plasticity
raw materials
shrinkage
sintering
workability

# Problems

1. Explain the importance of a flux in the processing of traditional ceramics.
2. Explain the effect of clay content on the processing of traditional ceramics.
3. How are the compounds used in modern technical ceramics prepared by comparison to those in traditional ceramics?
4. Explain the usefulness of an air classifier.
5. What are typical ceramic powder specifications?
6. What are important characteristics of the casting process? Name some products of this process.
7. What are the important features of the dry pressing process?
8. What is isostatic pressing?
9. What shapes are usually made by extrusion? What are the necessary characteristics in the powder mix?
10. What is sintering? What are important parameters associated with this process?

# 7

# *Composite Materials*

We can define composite material as **macroscopic combinations** of two or more distinct materials that have readily discernible interfaces between them. The combinations provide properties or behavior that the **constituent** materials do not possess individually. Although composites have been developed primarily for structural applications, as in the case of plastic and metal matrices, the composite concept also has been usefully applied to certain ceramic material systems to improve reliability and performance by deterring the catastrophic failure so characteristic of monolithic, brittle materials. Many composite materials comprise just two **phases**—the **matrix**, which is continuous and surrounds the second phase, the **reinforcement**.

Composites are often classified according to forms of the reinforcement, particulate, fiber, flake, and laminar composites. Fiber composites can be further divided into those containing discontinuous or continuous fibers.

A reinforcement is considered to be a particle if all of its dimensions are approximately equal. This applies to such geometries as spheres, flakes, short rods, and others of essentially equal axes. **Dispersion-strengthened** composites consist of a matrix containing very fine particles (0.01–0.1 $\mu$m in diameter). Particles may be added for purposes other than reinforcement as well; for example, the fillers described in Chapter 2 provide color, plasticity, reduced density, and so on.

Fiber-reinforced composites, shown in Figure 7.1, contain reinforcements whose lengths are much greater than their cross-sectional dimensions. These composites are considered to be **continuous** when the fiber lengths are about equivalent to the composite shape dimensions. Fiber dimensions range from less than one micrometer to well over a hundred micrometers. The processing methods employed for such composites often call for woven preforms that are infiltrated by the matrix material.

# 7.1  Dispersion-Strengthened Composites

Metals and metal alloys may be strengthened and hardened by the uniform dispersion of several volume percent of fine particles of a very hard inert material. Oxide particles are often used and strengthening occurs because the small dispersed particles hinder and halt the motion of dislocations through the metallic matrix under an applied load. In this way, plastic deformation is restricted such that yield strength and tensile strength are maintained, but ductility is reduced. For metals, dispersion strengthening is not as effective as precipitation hardening; however, the strengthening is retained at higher temperatures. Examples of such metal matrix systems include TD nickel, nickel metal reinforced by finely dispersed particles of thoria

**Figure 7.1**
Fiber composite structures:
(a) continuous and aligned,
(b) discontinuous and
aligned, (c) discontinuous
and randomly oriented

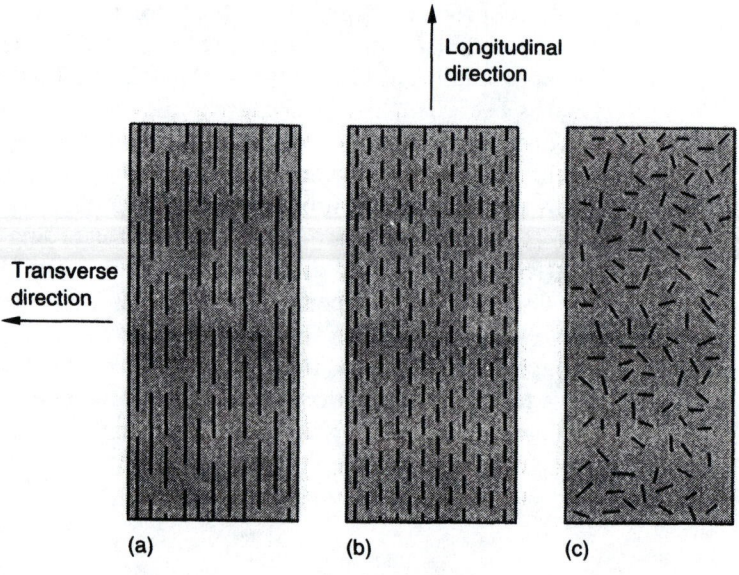

Longitudinal direction

Transverse direction

(a)          (b)          (c)

(ThO$_2$), and the aluminum–aluminum oxide system, where a very thin and adherent alumina coating forms on the surface of very small (0.1–0.2 $\mu$m) flakes of aluminum, which are then dispersed within an aluminum metal matrix. This material is called sintered aluminum powder (SAP). Although these materials have not found extensive use in engineering applications because of fabrication cost and property trade-off issues, thoriated tungsten is extensively used as the primary filament in electric light bulbs. Dispersed ThO$_2$ in the tungsten matrix acts as a deterrent to grain growth of the tungsten matrix, thereby maintaining tensile strength at the low-temperature start-up cycle where the filament is very brittle as well as at the high-temperature steady state condition.

# 7.2 *Particle-Reinforced Composites*

Because of their larger particle size by comparison to dispersed particles, these particulate systems may or may not reinforce matrix phases with which they are combined. In metallic matrices, these larger sized particles act to restrain movement of the matrix phase in the vicinity of the particle rather than interfere and halt dislocation mobility. The matrix then transmits some of the applied stress to the reinforcing particles and the degree of reinforcement depends on the strength of the bonding at the interface between the particle and the matrix.

Properties are influenced by particle size and distribution as well as by the volume concentration employed, with stiffness or elastic modulus increasing with increased particulate content (provided of course that the particulate has a higher elastic modulus than the matrix phase). Two mathematical equations express the elastic modulus of such particulate composites as a function of the volume fraction of each constituent phase for a two-phase system. These equations of the **rule of mixtures** type predict that the elastic modulus should fall between an upper value represented by

$$E_c = E_m V_m + E_p V_p$$

and a lower value given by

$$E_c = \frac{E_m E_p}{V_m E_p + V_p E_m}$$

where $E$ is the elastic modulus, $V$ is the volume fraction, and the subscripts $c$, $m$, and $p$ represent composite, matrix, and particulate phases, respectively.

Particulate composites have been used with all three material types (metals, polymers, and ceramics). Perhaps the most industrially important particulate composites are those belonging to the group known as the cemented carbides that are basically ceramic-particle metal-matrix combinations. In these compositions, titanium carbide (TiC) and/or tungsten carbide (WC) are embedded in a matrix of nickel-cobalt alloy and, in an appropriate shape, are extensively used for cutting tools and drill bits. The very hard carbide particles provide the cutting surfaces but, because they are brittle, cannot withstand the impact loads and stresses without early fracture and failure. Resistance to fracture is enhanced by their inclusion in a ductile metal matrix, which isolates the particles from one another, prevents particle-to-particle crack propagation, and thereby increases the **toughness,** or ability to absorb energy without failure. Large volume fractions of the carbide phases can be used (e.g., 80–90%) to maximize cutting performance. It is really extraordinary that the low volume concentration of metal matrix used not only provides the toughness needed for the composite, but also makes it possible to densify the material by providing a liquid phase at the sintering temperature. A photomicrograph of a cemented carbide tool composite is shown in Figure 7.2.

Polymeric materials are also reinforced with various particulates to improve overall properties. An outstanding example is the addition of carbon particles to rubber to enhance tensile strength and tear and abrasion resistance. Automobile tires, for example, contain in the range 15–30 volume percent of carbon black, whose particle size is in the range 20–50 nm.

# 7.3   *Fiber-Reinforced Composites* _____

When we consider composites, probably the first thing we think of is fiber composites, that is, fibrous materials embedded most likely in a plastic matrix. This is easy to understand because a main objective of fiber reinforcement is to provide a material with high strength and high elastic modulus with respect to the total weight. Polymer matrices satisfy the low specific gravity part of this need and many are readily combined with most strong fibers. We will look at polymeric matrix fiber composites, but here we must point out that many exciting developments are taking place in metal matrix and ceramic matrix composites.

**Figure 7.2**
Photomicrograph of a cemented
carbide (100×)
(From *Metals Handbook,*
ASM International, 1948)

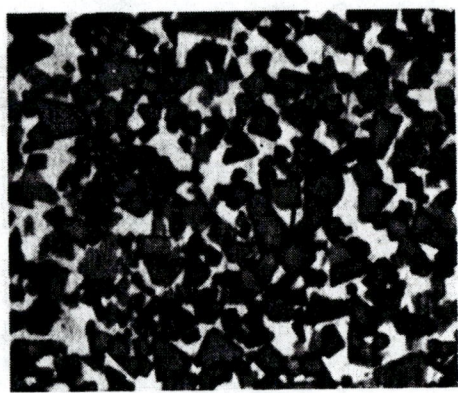

Fiber-reinforced composites were developed in response to the needs of the aerospace industry. Aluminum alloys, which provide high strength and fairly high elastic modulus or stiffness at low weight, have performed well and have been main-stay materials used in aircraft structures over the years. However, both fatigue and corrosion in aluminum alloys have produced problems that have been very costly to remedy. Fiber-reinforced composites were first developed in the 1940s. Fiberglass-reinforced plastics were used successfully in many applications, including filament-wound rocket motors. These materials were put to broader use in the 1950s and inexpensive fiberglass composites are used today in numerous consumer as well as aerospace products. Current availability of new and stronger fibers with properties superior to fiberglass has extended the performance range of fiber-reinforced com-posites, ensuring many future product developments.

# 7.3.1 Fiber Materials

Almost all high-strength, high elastic modulus materials fail by catastrophic propa-gation of flaws. A fiber of such a material, however, displays higher strength because of what is called the size effect. Tensile strength is basically dependent on a statisti-cal distribution of flaws, with larger forms of the same material having larger and more frequent flaws. They exhibit lower strength values than smaller cross-sectional forms. In addition, if equal volumes of fibrous and bulk material are compared, we would find that a crack due to a broken fiber would not easily propagate to cause the entire assemblage of fibers to fail, but failure would occur in bulk material from a similar flaw.

On the basis of diameter and character, fibers may be grouped into three dif-ferent classifications: whiskers, fibers, and wires. **Whiskers** are very small diameter single crystals that have large length-to-diameter ratios, for example, 20–50 $\mu$m in length and 0.1–1 $\mu$m in diameter. As a consequence of their small size and synthesis, they have a high degree of perfection for their exceptionally high strengths and, in this form, they are the strongest known materials. In spite of these high strengths, whiskers are not used extensively as a reinforcement medium because they are expensive, difficult to orient and classify into uniform diameters and lengths, diffi-cult to deagglomerate and distribute uniformly in a matrix phase, and difficult to bond to in many instances. Nonetheless, several applications where they are cur-rently employed involve metal and ceramic matrices to be described later in this chapter. Whisker materials include carbon (as the graphite structure), aluminum oxide (also known as sapphire whiskers), silicon carbide, and silicon nitride. Some of their characteristics are given in Table 7.1.

Materials that are classified as **fibers** are either polycrystalline or amorphous and have diameters larger than those of whiskers, that is, about 10 $\mu$m, and are pro-duced as continuous filaments and are wound on spools. These include such compo-sitions as glass, carbon, silicon carbide, aluminum oxide, boron, and polymer aramids. Two types of SiC are available, one of which results from the processing and controlled decomposition of polycarbosilane to yield continuous filaments of

**Table 7.1**
Characteristics of several fiber-reinforcement materials

| Material | Specific gravity | Tensile strength, psi × 10⁶ (MPa × 10³) | Specific strength, psi × 10⁶ | Modulus of elasticity, psi × 10⁶ (MPa × 10³) | Specific modulus, psi × 10⁶ |
|---|---|---|---|---|---|
| **Whiskers** | | | | | |
| Graphite | 2.2 | 3 (20) | 1.36 | 100 (690) | 45.5 |
| Silicon carbide | 3.2 | 3 (20) | 0.94 | 70 (480) | 22 |
| Silicon nitride | 3.2 | 2 (14) | 0.63 | 55 (380) | 17.2 |
| Aluminum oxide | 3.9 | 2–4 (14–28) | 0.5–1.0 | 60–80 (415–550) | 15.4–20.5 |
| **Fibers** | | | | | |
| Aramid (Kevlar 49) | 1.4 | 0.5 (3.5) | 0.36 | 19 (124) | 13.5 |
| E-glass | 2.5 | 0.5 (3.5) | 0.20 | 10.5 (72) | 4.2 |
| Carbon or Graphite | 1.8 | 0.25–0.80 (1.5–5.5) | 0.18–0.57 | 22–73 (150–500) | 15.7–52.1 |
| Aluminum oxide | 3.2 | 0.3 (2.1) | 0.09 | 25 (170) | 7.8 |
| Silicon carbide | 3.0 | 0.50 (3.9) | 0.17 | 62 (425) | 20.7 |
| **Metallic wires** | | | | | |
| High-carbon steel | 7.8 | 0.6 (4.1) | 0.08 | 30 (210) | 3.9 |
| Molybdenum | 10.2 | 0.2 (1.4) | 0.02 | 52 (360) | 5.1 |
| Tungsten | 19.3 | 0.62 (4.3) | 0.03 | 58 (400) | 3.0 |

small diameter, essentially amorphous SiC. The other process involves the chemical vapor deposition of SiC onto a carbon filament substrate to yield a fine-grained polycrystalline product approximately 125 $\mu$m in diameter and also of continuous length. The same manufacturer of this larger diameter SiC filament produces boron fiber in a similar cross-sectional diameter. The boron is chemically vapor deposited onto a moving tungsten filament to provide a continuous boron fiber. Data for these materials are also provided in Table 7.1.

    Glass fibers are produced by drawing monofilaments of glass from a furnace containing molten glass, coating the monofilaments with a polymer to "dull" any surface cracking, and gathering a large number of these filaments to form a strand of glass fibers, as depicted in Figure 7.3. The strands then are used to make glass fiber yarns, or rovings, that consist of a collection of bundles of continuous filaments. Considering the data of Table 7.1, it can be seen that glass fibers have the lowest elastic modulus, if not the lowest tensile strength, compared to the other fibers. However, because of their much lower cost and ready availability, glass fibers are by far the most commonly used reinforcing fibers for plastics.

Carbon fibers have a combination of very high strength, low density, and high elastic modulus. These properties make the use of carbon fiber–plastic composite materials especially attractive for aerospace applications. Carbon fibers are produced mainly from two sources, polyacrylonitrile (PAN) and pitch, which are called **precursors**. In general, carbon fibers are produced from PAN precursor fibers by three processing stages: (1) stabilization, (2) carbonization, and (3) graphitization. In the stabilization stage, the PAN fibers are first stretched to align the fibrillar networks within each fiber parallel to the fiber axis. Then they are oxidized in air at about 200–220°C while held in tension in order to provide cross-linking between the fibrils to avoid melting at the next stage. In the second stage, carbonization, the cross-linked fibrils are pyrolated (heated) until they become transformed into carbon fibers by the elimination of O, H, and N from the precursor fiber. This carbonization treatment is car-

**Figure 7.3**
Fiberglass forming process (From W. F. Smith, *Principles of Materials Science and Engineering*, 2nd ed., McGraw-Hill Publishing Company, 1990.)

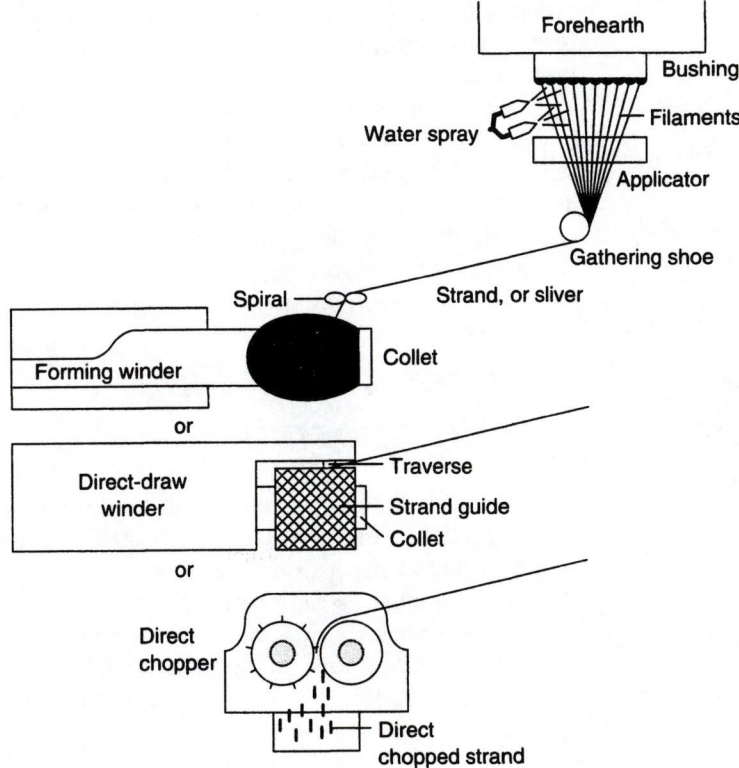

**Figure 7.4**
Kevlar polymer chain repeating unit, or mer

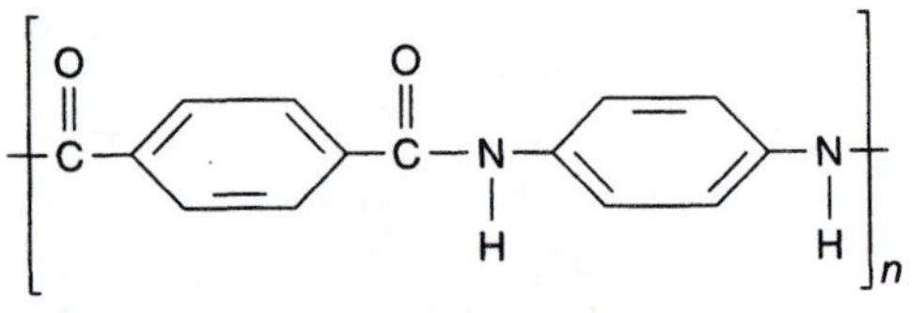

ried out in an inert atmosphere in the range of 1000–1500°C. During the carbonization process, graphitelike fibrils, or ribbons, are formed within each fiber and provide the great increase in tensile strength observed. The third stage involves complete conversion of the fiber to oriented graphite crystal form by heating the fiber while under tension to temperatures above 2000°C. This graphitization procedure can raise the elastic modulus to levels exceeding $50 \times 10^6$ psi (250 GPa). Carbon fibers produced from PAN precursor material have tensile strengths that range from about 450 to 650 ksi (3.10 to 4.45 GPa) and elastic moduli that range from 28 to $50 \times 10^6$ psi (193 to 250 GPa). The density of the carbonized and graphitized PAN fibers is usually about 1.8 g/cm³ with fiber diameters of 7 $\mu$m to 10 $\mu$m.

Aramid fibers were introduced commercially by the DuPont Corporation in 1972 under the trade name Kevlar and currently these are available in two commercial types, Kevlar 29 and Kevlar 49. Kevlar 29 is a low-density, high-strength fiber designed for such applications as ropes, cables, and even ballistic armor. The properties of Kevlar 49 make its fibers useful as reinforcement for plastics in composites for the aerospace, marine, and automotive industries as well as many other applications.

The chemical repeating unit of the Kevlar polymer chain is that of an aromatic polyamide mentioned in Chapter 2 and repeated here in Figure 7.4. The aromatic ring structure gives high rigidity to the polymer chains, causing them to have a rodlike structure. Strong covalent bonding in the polymer chains provides the strength and high elastic modulus character. Kevlar aramid is used for **high-performance** composite applications where light weight, high strength, stiffness, and impact and fatigue resistance are important.

# 7.3.2 Polymer-Matrix Fiber-Reinforced Composites

The purpose of matrix polymers is to bind fibers together by virtue of their adhesive characteristics so that mechanical loads may be transferred from the weak matrices to the higher strength fibers. In these composites, binding strength between fibers and matrices must be high to minimize fiber pullout, which causes premature failure. The matrices can also serve to protect the fibers from handling damage and environmental degradation, and matrix selection generally determines overall service temperature limitations of a composite as well as processing conditions during fabrication.

Polyester resins are the most commonly used matrices for fiberglass-base composites. These thermosetting resins offer a combination of low cost, versatility in many processes, and good property performance.

## Sample Problem 7.1

A continuous unidirectionally aligned glass fiber–reinforced composite consists of 45 volume percent of glass fibers having a modulus of elasticity of $10 \times 10^6$ psi ($69 \times 10^3$ MPa) and 55 volume percent of a polyester resin that has a modulus of $5 \times 10^5$ psi ($3.5 \times 10^3$ MPa).

    a.  Compute the modulus of elasticity of this composite in the fiber aligned direction.

    b.  If the cross-sectional area is 0.5 in.² (323 mm²) and a stress of 10,000 psi (69 MPa) is applied in this direction, calculate the load carried by each of the fiber and matrix phases.

    c.  Determine the strain that is sustained by each phase when the stress in part b is applied.

    d.  Assuming tensile strengths of 500,000 psi ($3.5 \times 10^3$ MPa) for glass fibers and 10,000 psi (69 MPa) for polyester resin, determine the tensile strength of this fiber composite in the fiber aligned direction.

### Solution

Basic relations for longitudinally reinforced fiber composites with well-bonded axially aligned fibers are

$$E_c = E_f V_f + E_m V_m$$
$$\sigma_c = \sigma_f V_f + \sigma_m V_m$$
$$\epsilon_c = \epsilon_f = \epsilon_m$$

where $E$ is the elastic modulus, $\sigma$ is the tensile strength, and $\epsilon$ is the strain. The subscripts $c$, $f$, and $m$ represent composite, fiber, and matrix (polymer), respectively.

    a.  $E_c = 10 \times 10^6$ psi $(0.45) + 5 \times 10^5$ psi $(0.55)$
       $= 4.8 \times 10^6$ psi $(= 3.3 \times 10^4$ MPa)

    b.  To solve this section of the problem, we must find the ratio of fiber load to matrix load. Since the strains are equal when stress is applied longitudinally,

$$\epsilon_c = \epsilon_f = \epsilon_m$$

Since stress / strain = modulus $(\sigma/\epsilon = E)$ and load / area = stress $(F/A = \sigma)$, we have

$$\epsilon_c = \frac{F_c/A_c}{E_c} = \frac{F_f/A_f}{E_f} = \frac{F_m/A_m}{E_m}$$

Transposing, we obtain

$$\frac{F_f}{F_m} = \frac{E_f(A_f)}{E_m(A_m)} = \frac{10 \times 10^6 \,(0.45)}{5 \times 10^5 \,(0.55)} = \frac{16.4}{1}$$

$$F_f = 16.4 F_m$$

where $F_f$ is load on fiber and $F_m$ is load on matrix.

Now the total force sustained by the composite $(F_c)$ may be calculated from the applied stress $\sigma$ and total composite cross-sectional area $A_c$ according to

$$F_c = A_c\sigma_c = (0.5 \text{ in.}^2)(10,000 \text{ psi})$$
$$= 5000 \text{ lb of force } (22,143 \text{ N})$$

However, this total load is the sum of the loads carried by fiber and matrix phases, that is,

$$F_c = F_f + F_m = 5000 \text{ lb}$$

Substitution from preceding calculations yields

$$16.4 F_m + F_m = 5000 \text{ lb}$$
$$F_m = 287 \text{ lb } (870 \text{ N})$$
$$\text{and } F_f = F_c - F_m$$
$$= 5000 \text{ lb} - 287 \text{ lb} = 4713 \text{ lb } (20,872 \text{ N})$$

Thus the fiber phase supports most of the applied load.

c. The stress for both fiber and matrix phases must first be calculated. Then by using the elastic modulus for each (from part a), the strain values may be determined. For stress calculations, phase cross-sectional areas are necessary and for continuous equal length fibers, the volume fraction is equal to the area, not a real fraction, so that

$$A_m = V_m A_c = (0.55)(0.5 \text{ in.}^2) = 0.275 \text{ in.}^2 \,(177 \text{ mm}^2)$$
$$\text{and}$$
$$A_f = V_f A_c = (0.45)(0.5 \text{ in.}^2) = 0.225 \text{ in.}^2 \,(145 \text{ mm}^2)$$

Thus $\sigma_m = \dfrac{F_m}{A_m} = \dfrac{287 \text{ lb}}{0.275 \text{ in.}^2} = 1044 \text{ psi } (7.2 \text{ MPa})$

$$\sigma_f = \frac{F_f}{A_f} = \frac{4713 \text{ lb}}{0.225 \text{ in.}^2} = 20,947 \text{ psi } (144.5 \text{ MPa})$$

Finally, strains are calculated as

$$\epsilon_m = \frac{\sigma_m}{E_m} = \frac{1044 \text{ psi}}{5 \times 10^5 \text{ psi}} = 0.0021$$

$$\epsilon_f = \frac{\sigma_f}{E_f} = \frac{20,947 \text{ psi}}{10 \times 10^6 \text{ psi}} = 0.0021$$

Therefore, strains for both matrix and fiber phases are identical, which they should be.

d. For tensile strength, we have

$$\sigma_c = \sigma_f V_f + \sigma_m V_m$$
$$= 500{,}000(0.45) + 10{,}000(0.55)$$
$$= 230{,}500 \text{ psi } (1586 \text{ MPa})$$

The importance of adding glass fiber to increase strength is clearly shown.

---

Even though reinforcement efficiency is lower for discontinuous than for continuous fibers, **discontinuous** (short length) fiber composites are becoming important in the commercial market because of lower processing costs. Chopped glass fibers are used extensively in polyester matrices. These short fiber composites can be produced to have tensile strengths that approach 50% of their continuous fiber counterparts.

Problems with **adhesion** to carbon and aramid fibers have discouraged the development of polyester composites that use these fibers. Although there are high-performance fiberglass composites in military and aerospace structures, the relatively poor performance of advanced composites of polyester when used with fibers other than glass and the comparatively large cure shrinkage of these resins have generally restricted such composites to lower performance applications.

When property requirements justify the additional costs, epoxies and other resins are used in high-performance commercial applications such as sporting goods (tennis rackets, fishing rods), printed circuit boards, and chemical piping. Epoxy resins are used far more than all other matrices in advanced structural applications. Although epoxies are sensitive to moisture in both their cured and uncured states, they are generally superior to polyesters in resisting moisture and other environmental influences and offer better mechanical properties at acceptable cost. A substantial data base exists for epoxy resins because both the U.S. Air Force and U.S. Navy have been flying aircraft with epoxy matrix structural components since 1972 and the in-service experience with these components has been very satisfactory.

Currently, carbon fiber–reinforced epoxy matrices are the most widely used composites for aerospace structural and other high-performance applications. The main advantage of carbon fibers is that they have very high strength coupled with high elastic moduli and low density. Carbon fiber–epoxy matrix composites have replaced much of the aluminum used in modern aircraft structures where weight reduction is so important. Figure 7.5 shows the exceptional fatigue properties of **unidirectional** carbon (graphite) fiber–epoxy material as compared to those of aluminum alloy 2024-T3 and other composite materials.

In engineering-designed structures, fiber-polymer matrix material is laminated using precast **fabric plies** of fibers in desired orientations so that tailor-made strength requirements are met. Figure 7.6 compares schematics of unidirectional and **multidirectional** plies for a composite laminate.

**Figure 7.5**
Fatigue properties of some unidirectional fiber composites compared to 2024-T3 aluminum alloy
(From W. F. Smith, *Principles of Materials Science and Engineering*, 2nd ed., McGraw-Hill Publishing Company, 1990.)

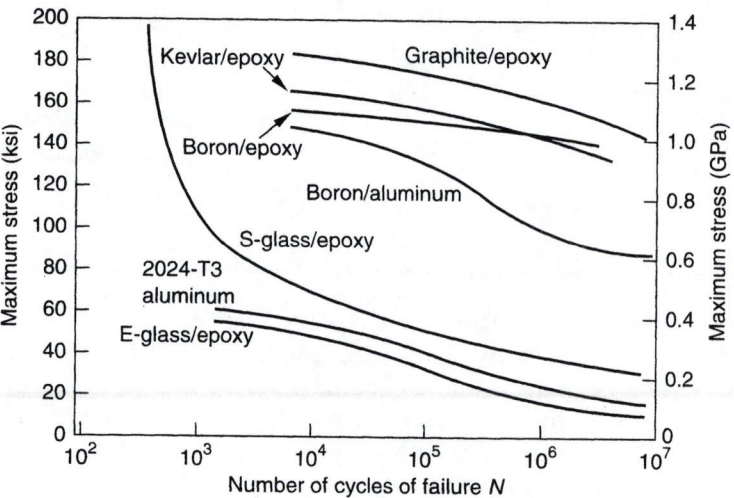

Although the preceding discussion has centered on single-fiber single-matrix combinations, we should be aware that fiber-reinforced matrix structures can take many hybrid forms. This is particularly true in the bonded abrasives industry. Case Study 7.1 bears out the importance of selecting the correct composite structure for the job at hand.

Effort is ongoing to extend the service temperature limit of 120°C for epoxy resin systems by investigating other formulations. Some high temperature resins possess many of the same desirable features as epoxies, such as fair handleability, relative ease of processing, and excellent composite bonding behavior. They are also superior in extending the safe in-service temperature to about 220°C; however, they have a lower elongation to failure than epoxies and are quite brittle. Work is still continuing to improve this system.

Polyimide resins are available with a maximum in-service temperature of about 260°C. These thermosetting resins, unlike the others, normally cure by a condensation reaction that releases volatiles. This creates a problem because the released volatiles produce undesired voids in the composite. Recent effort has been directed at reducing this problem by using an addition reaction during curing that does not release volatiles. Although these resins will produce low void content composite parts, they are also brittle, with poor impact resistance.

The attempts to improve thermosetting resins continue with major efforts focused on raising temperature stability and decreasing brittleness, which would improve impact resistance. The dual goal of improving temperature resistance and impact resistance of composite matrices has led to the development and limited use of new high-temperature thermoplastic resin matrices. These materials are very different from the well-known thermoplastics, such as polyethylene, polyvinyl chloride, and polystyrene, that are commonly used as plastic bags, piping, and tableware and have little resistance to elevated temperature.

These high-temperature thermoplastics are tougher and offer the potential of improved temperature resistance over epoxies. They also exhibit higher strains to

**Figure 7.6**
Unidirectional and multidirectional fabric ply laminate composites
(From W. F. Smith, *Principles of Materials Science and Engineering*, 2nd ed., McGraw-Hill Publishing Company, 1990.)

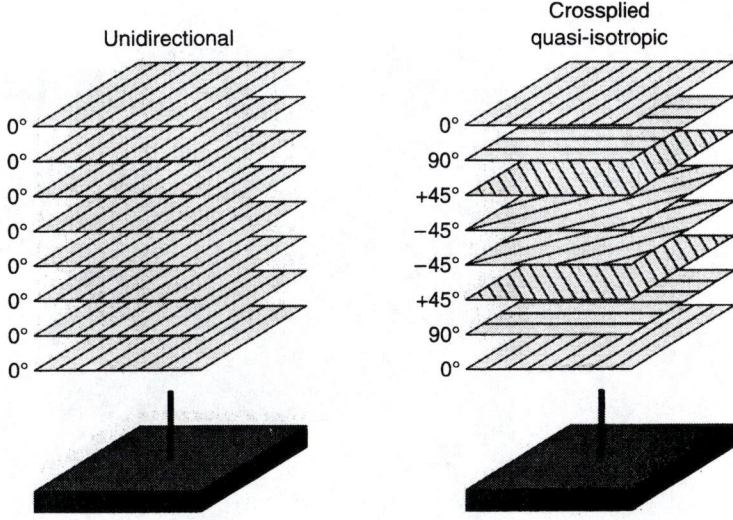

failure that should improve impact resistance. These materials include such resins as polyetherketone (PEEK), polyphenylene sulfide (PPS), and polyetherimide (PEI), all of which maintain thermoplastic character after processing. The fabrication procedures necessary for low-cost manufacture of thermoplastic matrix composites are being thoroughly investigated and major efforts are concerned with determining and understanding the mechanical properties and behavior of fabricated composites.

## Case Study 7.1

### Selecting a Composite Abrasive Finishing Wheel

Harrell Die Casting, Inc., produces zinc alloy die castings for a variety of applications. These castings must be finished, that is, gates and parting lines must be cut and/or smoothened by abrasive grinding. Flexible grinding wheels that use a cotton-fiber, resin-bonded $Al_2O_3$ abrasive have always been purchased from Allied Distributors. The standard 7-in. wheel was rated at 7750 RPM maximum speed, quite acceptable for the 6500 RPM grinder in use.

When a replacement order was placed, Allied Distributors did not have the 7-in. wheels in stock. Consequently, delivery would be delayed at least 4 weeks, which would force the Harrell job orders to be delivered late, quite unacceptable to Harrell's customer. The clerk at Allied Distributors located some 9-in. wheels that, although not labeled, appeared to be the same cotton-fiber, resin-bonded $Al_2O_3$ abrasive wheel. Harrell Die Casting accepted the substitute wheels, to be used only until the correct wheels could be delivered.

This decision proved to be a poor one because the safeguard for the grinder had to be removed in order to accommodate the larger wheel and the wheel broke up in the first finishing operation attempted. Fortunately, no one was in the path of the exploding wheel. Nevertheless, production commitments could not be satisfied.

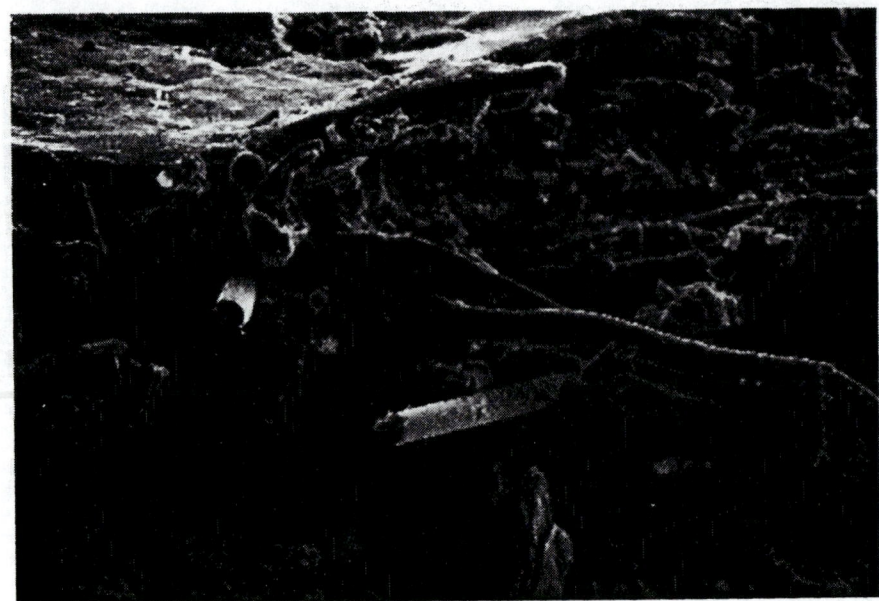

**Figure 7.7**
SEM micrograph of fracture surface of cotton-fiber, resin-bonded $Al_2O_3$ (400×)

Before Harrell confronted Allied Distributors, they had a section of the broken wheel examined in a scanning electron microscope (SEM). To their surprise, they found the composite wheel to have not only cotton fibers but also glass fibers. The glass fibers provide strength against traditional rotational forces, but for a flexible wheel where bending forces are also encountered, the fiberglass is brittle. Figure 7.7 illustrates two glass fibers that display brittle fracture and several unbroken helically wound cotton fibers.

After a meeting between Harrell and Allied Distributors, the manufacturer of the wheels was called. New cotton-fiber, resin-bonded $Al_2O_3$ abrasive 7-in. wheels were delivered by Federal Express the next morning. (The safeguard, of course, was replaced and delivery commitments were honored, but not without requiring employees to work overtime.)

## 7.3.3 Metal-Matrix Fiber-Reinforced Composites

Metal matrix composite systems are emerging as candidate materials to go beyond the plastic composites in terms of service temperatures. These composites consist of a metal base that is reinforced with one or more constituents, such as continuous boron, silicon carbide, graphite, alumina fibers, or ceramic materials in whisker or particle form. In the case of the continuous fiber–reinforced composites, the fiber is the dominating constituent and the metal matrix serves as the medium for transmitting the load to the reinforcing fiber. For effective load transfer to occur, the fiber modulus must be much higher than the matrix modulus and each of the constituent materials must have moduli sufficiently greater than many potential metal matrices.

Boron was the first high-strength, high elastic modulus reinforcing fiber to be used in metal matrix composite applications. Both aluminum and titanium were investigated as matrices and the production of boron-reinforced aluminum composites has been moderately successful, with specific applications in the aerospace industry. The effort with titanium has been curtailed, however, because of severe processing environments that degraded the strength and stiffness of the boron filaments.

The primary advantage of the boron-reinforced metal matrix over its boron-epoxy counterpart is the maximum operating temperature to which the former can be exposed. For example, boron fiber–reinforced aluminum offers useful mechanical strength properties up to about 500°C, whereas an equivalent boron-epoxy composite is limited to about 190°C. Figure 7.8 compares the specific tensile strengths of several materials as a function of temperature and shows the particular advantage of boron-aluminum. Applications for which boron-reinforced aluminum composites have been evaluated and used include jet engine fan blades, aircraft wing skins, structural supports, landing gear components, bicycle frames, and golf club shafts. Examples are provided in Figures 7.9 and 7.10.

**Figure 7.8**
Strength of axially reinforced boron composites versus metals
(From *Engineered Materials Handbook*, Vol. 1: *Composites*, ASM International, 1987.)

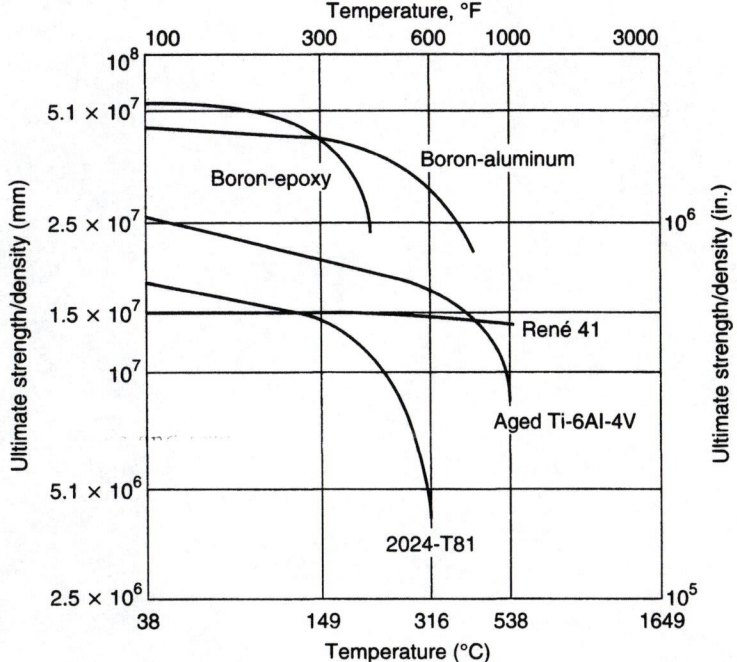

**Figure 7.9**
Schematic of mid-fuselage
structure of space shuttle
(From *Engineered Materials
Handbook*, Vol. 1: *Composites*, ASM International,
1987.)

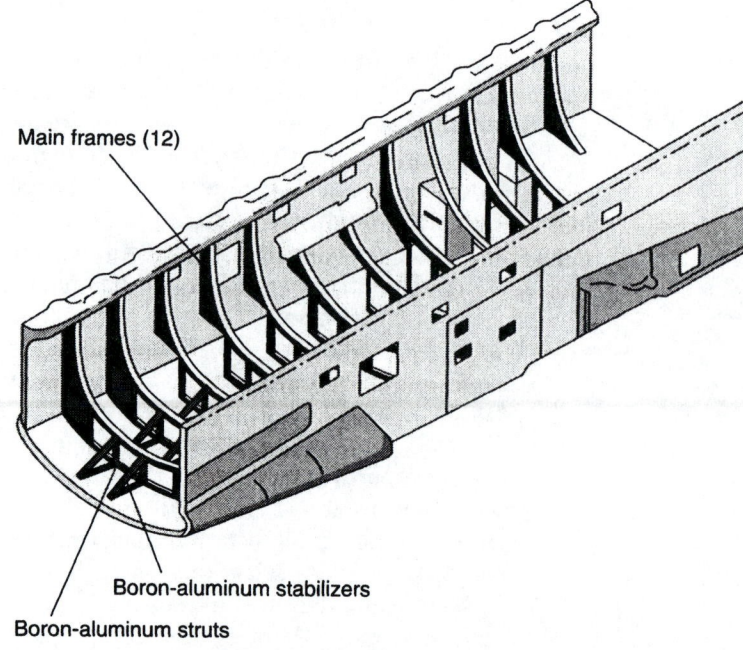

Main frames (12)

Boron-aluminum stabilizers

Boron-aluminum struts

**Figure 7.10**
Boron fiber–aluminum
composite bicycle frame
(From *Engineered Materials
Handbook*, Vol. 1:*Composites*,
ASM International, 1987.)

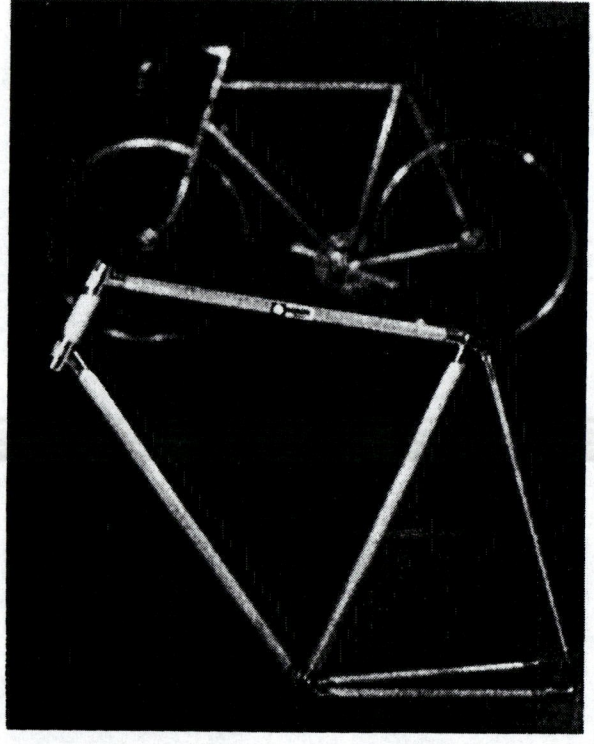

Temperature limitations in both the use and the processing of boron fiber composites have more recently led to the development of metal matrix composites based on silicon carbide fibers. Silicon carbide fiber–reinforced metals can be produced more easily because the SiC fibers readily bond to metal matrices and resist strength degradation during high-temperature processing. Besides aluminum, SiC fibers have successfully reinforced titanium, which is under consideration as an advanced fan blade material in the compressors of aircraft turbine engines. The silicon carbide raises the operating temperature, improving fuel efficiency while lowering compressor weight.

Discontinuously reinforced metal matrix composites exhibit a blend of reinforcement and matrix properties. The reinforcement can be ultra-high-strength whiskers, short or chopped fibers, or even particles. Early work on the use of whiskers was performed in the 1960s with aluminum oxide and abandoned shortly afterward because the whisker costs were high and strengths achieved were lower than expected because of bonding difficulties between the alumina whiskers and metal matrices. The more recent availability of SiC whiskers at lower cost and improved performance as a reinforcement for aluminum alloys has renewed the interest in discontinuous fiber-reinforced metals. Potential applications include automobile engine pistons that have greater resistance to wear and corrosion.

# 7.3.4 *Other Metal Matrix Composites*

There are a number of composites that are unique because they are designed for specific reasons that are nonstructural. For example, one method that has been investigated for producing composites is the directional solidification of eutectic composites. Although this method is impractical because of expense, many interesting microstructures with unique characteristics have been developed. One such composite alloy is a cobalt-niobium eutectic where lamellar $Co_3Nb$ in a cobalt matrix was formed by directional solidification. Significant mechanical strengthening was exhibited, but the magnetic properties were reduced by the amount of niobium in solid solution. This was counteracted by adding iron to the composition, all iron being dissolved in the cobalt matrix, thus increasing the magnetic saturation moment and reducing the coercive force. Figure 7.11 illustrates the microstructure of the transverse and longitudinal sections of a ternary eutectic alloy containing about 17 atomic percent Fe.

A much more practical metal matrix composite that has commercial application is the multifilamentary superconducting wire used in high-field magnets for magnetic resonance imaging (MRI) equipment and high-speed magnetic levitation. Although high-temperature 1-2-3 superconductors are today's research phenomena in the field of superconductivity, the only highly utilized superconducting material is an alloy of niobium and titanium. When superconductors are below the critical temperature ($T_c$), have current density below a critical value ($J_c$), and are in a magnetic field below a critical value ($H_c$), they display zero resistance. Therefore, **wires** can be wound into high magnetic field solenoids. But all superconductors "go normal" if

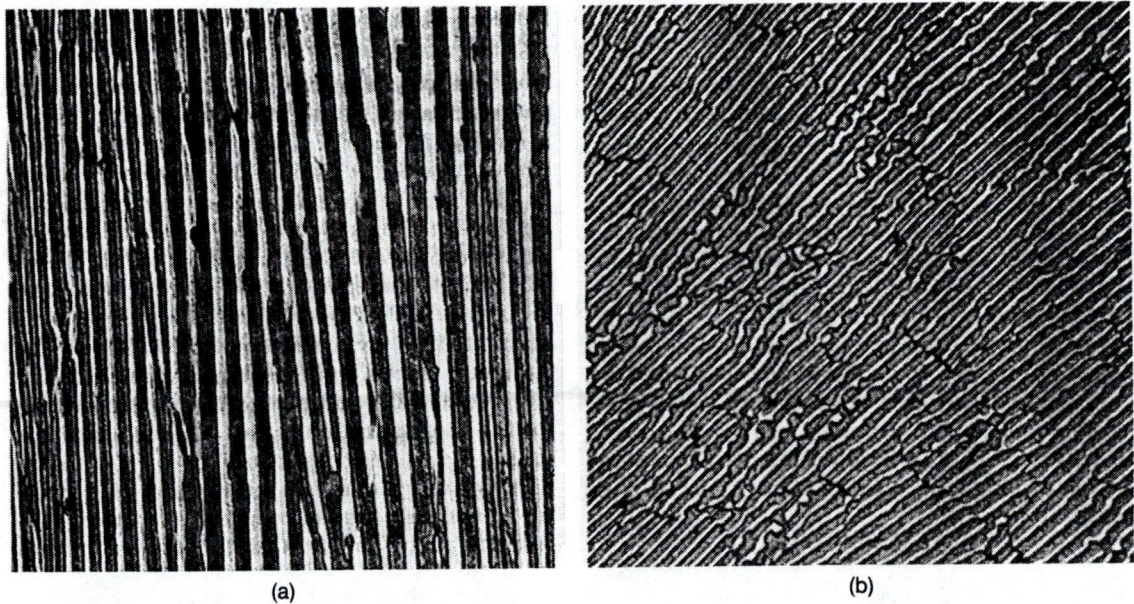

(a)                                                    (b)

**Figure 7.11**
Directionally solidified ternary eutectic Co-Nb-Fe alloys (500×): (a) 17 atomic percent
Fe, longitudinal section, (b) 17 atomic percent Fe, transverse section

any of these values are exceeded. In the normal (nonsuperconducting) condition, they have high electrical resistance and can overheat and melt, irreparably damaging the utility of the magnet.

Such a delicate balance is avoided by the concept of a composite of superconducting alloy filaments in a matrix of very pure copper, an excellent conductor, that will dissipate any heat generated by the magnet going normal during operation. Thus the magnet will not be destroyed. Manufacture of a fine filamentary composite would also be beneficial because cold work increases the current-carrying capacity. To maximize the amount of cold work, however, very rigid processing controls are necessary. The procedures that have been adopted in the industry include extensive component cleaning, stacking of annealed alloy rods into very pure copper tubes, and placement of the "bundled" tubes into a very pure copper can, which is welded and extruded into rod. The extrusion conditions are critical because the superconducting alloy cannot be annealed by adiabatic heating during the extrusion deformation; the filaments must remain below the recrystallization temperature to ensure work hardening. During wire drawing, a precipitation heat treatment also is required and twisting of the final wires is necessary to avoid another deleterious phenomenon called *flux jumping* in the final configuration. An example of the final composite wire that is produced in this manner is shown in Figure 7.12.

**Figure 7.12**
Multifilamentary superconducting composite with copper matrix and niobium-titanium alloy filaments (200x)

# 7.3.5 Ceramic-Matrix Fiber-Reinforced Composites

In contrast to the plastic and metal matrix composites we have discussed, where strength and stiffness enhancement were primary objectives for developing these materials, ceramic composites have been developed largely for other purposes because ceramics are inherently very stiff and strong. These purposes include improved resistance to thermal shock failure, more reliable tensile strength properties, and enhanced toughness and impact resistance. Many successful ceramic composites were developed to fulfill the needs and requirements of specific applications. Case Study 7.2 demonstrates the usefulness of a composite approach as a solution to a materials problem.

---

## Case Study 7.2

### Solid-Propellant Rocket Motor Nozzle Development

In the development of solid-propellant rocket motors, the selection and fabrication of the material to be used as the throat constriction or rocket nozzle is critical. Unlike liquid fuel motors that are cooled by the fuels themselves, solid-propellant nozzles can reach surface temperatures of 5000°C or more for short periods of time, usually less than a few minutes. Tungsten metal was the first material to be employed for this application in the late 1950s and early 1960s because of its very high melting point (3300°C) and good resistance to erosion, but its use was discontinued, particularly for large motors, because it was too heavy. Tungsten has one of the highest elemental densities at 19.3 gm/cc, and its replacement with low-density ceramics was desirable. SiC was first selected as a **monolithic** material (i.e., single-phase macroscopic) but somewhat modest temperature gradients caused thermal stress failure. The manufacturers of the SiC component tried to improve the thermal resistance by mixing in graphite particles with the SiC so that a nozzle part would have a matrix of SiC with a dispersion of graphite (between 10 and 20 volume percent). For some smaller motors, this material system worked well because thermal stress cracks that would develop in service would be blocked and virtually stopped at the interface between the graphite particles and the SiC matrix. The graphite acted as a barrier to crack propagation.

As the demands became more stringent on nozzles because of higher temperature propellants and increasing motor size, attempts were made to use bulk graphite materials. These materials were rather porous, with low modulus and strength properties but with high thermal conductivity. They eroded too quickly in use and were eventually replaced in the 1970s by the graphite fiber composite known as 3D carbon-carbon composite, a three-dimensional graphite fiber with a carbon graphite matrix. Also in the 1970s, smaller samples of these composites were being employed as nosecones in hypersonic missiles. The availability of graphite fibers at the time combined with textile weaving capability made the fiber preform construction possible. Techniques had to be developed to introduce a carbon matrix and this was accomplished by infiltrating with various resins and then decomposing the resins under controlled atmosphere to yield a carbon char that served as the matrix, whereby the preform was more rigid and porosity was reduced to low levels. This was achieved by several iterative cycles of infiltration and charring to reach the range of 90% of theoretical density levels. These carbon-carbon composites were an outstanding success and continue to be used today. The composites exhibit high toughness for a brittle composite. Any cracks that may develop in service are deflected at fiber matrix interfaces and the crack energy is usually dissipated in the surrounding remnant porosity. This was truly a remarkable materials development and carbon-fiber carbon-matrix composites are used as the nose cap and leading edges of wings, rudder, and tail assembly of the U.S. space shuttle vehicle. Attempts are also being made currently to develop oxidation-resistant coatings for this material system because of its potential as a weight-saving, higher temperature, load-bearing structural component in advanced turbine engines.

---

The idea of employing brittle fibers in a multidirectional configuration that are held together by a brittle matrix has wide appeal and is an area of continuing development. Figure 7.13 shows the microstructure of a fracture tough, thermal shock resistant fused silica fiber composite that is employed as a radar window in hypersonic vehicles. It must withstand high mechanical shock loads, be resistant to particle impact, and survive severe thermal stress conditions while transmitting radar sig-

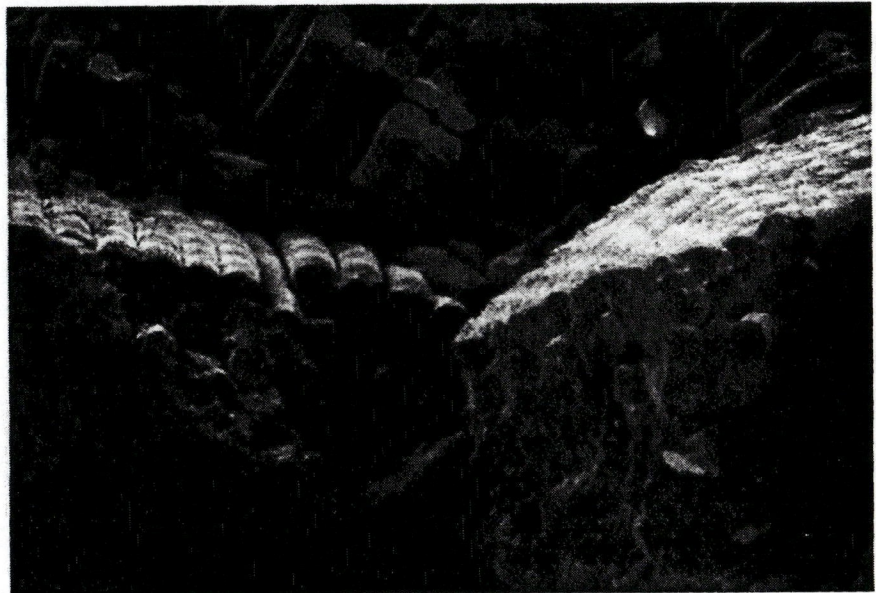

**Figure 7.13**
SEM micrograph of fused silica fiber composite used in radar window (500x)

nals. Ordered monolithic fused silica would not survive under these conditions but the composite approach provided the solution.

Currently under development are silicon carbide fiber-silicon carbide matrix composites that are fracture tough and also have good oxidation resistance. These composites do not have particularly high strength because of poor bonding between matrix and fiber but they offer greater reliability than monolithic ceramics that are more flow sensitive and more prone to catastrophic failure.

# *Summary*

Composite materials have emerged as very important structural materials, particularly where weight reduction and high performance are important objectives. Plastic matrix composites containing high-strength ceramic fibers, including graphite, dominate the composites used at low temperatures. First used in the aerospace industry, their application to ground-based transportation systems, commercial products, and sporting goods has become extensive. The low-modulus plastic matrices coupled with the high-strength, high elastic modulus fibers and the ability to position fibers in desired orientations leads to tailor-made materials with managed properties. Metal matrix composites offer the potential for higher temperature use and are still in various stages of development. A few ceramic matrix fiber composites have been notable for their success as high-performance aerospace materials. While their application in aerospace vehicles continues, attempts are being made to extend their application to other commercial activities.

# Terms to Remember

adhesion

constituents

continuous

discontinuous

dispersion strengthened

fabric ply

fibers

high performance

macroscopic combinations

matrix

monolithic

multidirectional

phases

precursor

reinforcement

rule of mixtures

toughness

unidirectional

whiskers

wires

# Problems

1. Define a composite material.
2. What are dispersion-strengthened composites?
3. Describe the size effect that occurs in tensile strength properties.
4. Describe the structural features that make aramid fibers strong.
5. Using the properties provided in Sample Problem 7.1 in the text, calculate the volume percent of glass fibers required to prepare a unidirectionally reinforced polyester resin composite with an elastic modulus of 8,000,000 psi.
6. What are the differences between whiskers and fibers?
7. What is a very important application of carbon fiber–reinforced epoxy resin composites and why has this composite been selected.
8. What would be an important advantage of metal-matrix fiber-reinforced composites over plastic matrix composites?
9. What is attractive about ceramic matrix-fiber reinforced composites?
10. Discuss the applications of ceramic matrix-fiber composites.

# 8

# *Processing of Fiber-Reinforced Composite Materials*

In this chapter, we will concentrate on fiber-reinforced materials because of the accelerated use of these materials in response to market needs. Many processes are available for the fabrication of composite materials and the methods we choose depend largely on the type of matrix selected, the shape to be fabricated, and on the fiber selection, concentration, and orientation.

## *8.1 Processing of Plastic Matrix Composites*

### *8.1.1 Lay-up Processes*

Prepreg tapes of continuous **fiber reinforcement** in uncured matrix resin are the most widely used preforms for **plastic matrix composites** made by the **lay-up**

**process** for structural applications. Use of tapes ensures uniform fiber placement and lower processing costs. Prepreg **tape** is a series of parallel fiber-reinforcing tows impregnated with a matrix resin (a *tow* is fiber spun from short lengths of filaments).

Fiber is typically converted into a prepreg by bringing a number of spooled fiber tows into a **collimated** form, as shown in Figure 8.1. The **prepregging** operation consists of heating a matrix resin to obtain low viscosity and creating a well-dispensed fiber-resin mass. The amount of fiber is controlled by the number of tows brought into the prepreg line, and the resin can be cast onto substrate paper either on the prepreg line or in a separate filming operation to obtain a desired fiber-to-resin ratio. The prepreg is calendered to obtain a uniform thickness and to close fiber gaps before being wound on a core. Substrate paper is ordinarily left between layers of tape and serves as a releasing film. The finished tape product is wound on a spool, interleaved with paper. Figure 8.2 shows a typical spool of graphite-epoxy prepreg tape that is available in a wide variety of widths, thicknesses, and package sizes. Table 8.1 lists typical ranges of tape product dimensions. We use narrow prepreg tape (75 mm, or 3 in.) as compared to wide tapes to minimize material loss to below 10%, but this savings can be offset by increased labor cost so manufacturing engineers must balance these costs for each component. The use of unidirectional tape in manufacturing processes falls into three major lay-up categories: (1) hand lay-up, (2) machine-cut patterns that are laid up by hand, and (3) automatic machine lay-up.

Historically, tapes have been used primarily in a hand lay-up procedure whereby an operator cuts lengths of tape and places them onto a tool surface, building the desired thickness and fiber ply orientation to form a laminate. After the lam-

**Figure 8.1**
Typical prepreg machine
(From *Engineered Materials Handbook*, Vol. 1: *Composites*, ASM International, 1987.)

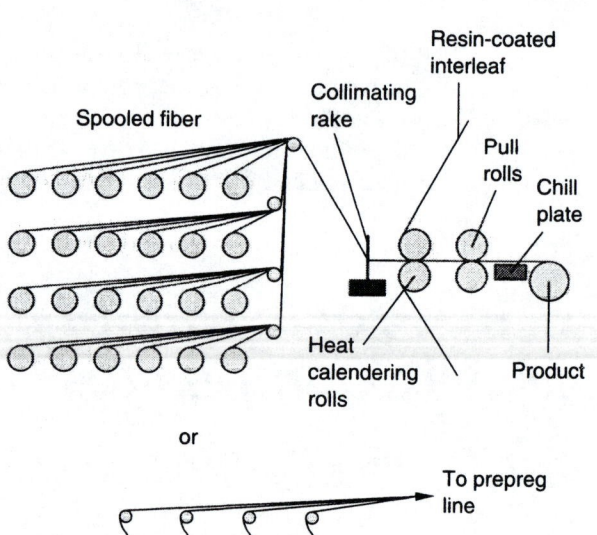

**Figure 8.2**
Graphite-epoxy tape
(From *Engineered Materials
Handbook*, Vol. 1: *Composites*,
ASM International, 1987.)

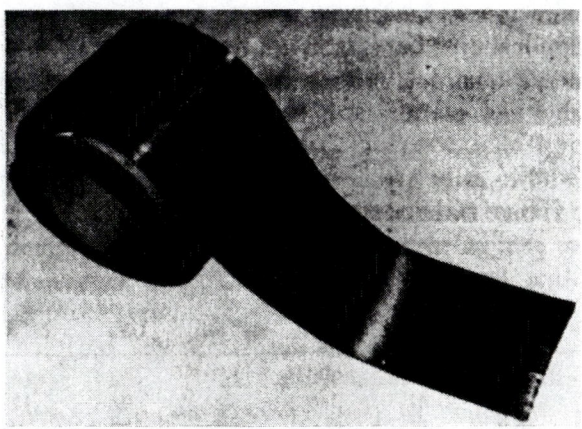

inate is completed, the *tooling* and attached laminate are bagged and a vacuum is applied to remove entrapped air. The whole setup is then put into an autoclave for final curing of the resin. Curing conditions vary for each material, but for carbon-epoxy composites, 190°C and 100 psi of applied pressure are usually adequate. On removal from the autoclave, the part is stripped from its tooling and is ready for any finishing operations. More recent technology uses machine-cut patterns that are then laid up by hand. This method involves higher investment in machinery, but increases production and reduces operator error in lay-up.

The latest technology uses controlled automatic tape-laying machines that are programmed to lay down plies of tape in desired patterns with consistent lay-down pressures and ply-to-ply separations. Modern automatic tape layers can handle only limited tape widths and simple tool contours, but soon will be available to fabricate large, heavily contoured parts. Figure 8.3 shows a carbon fiber–epoxy laminate of an AV-8B wing section and tooling being put into an autoclave for curing. This lay-up process is used mainly in the aerospace industry where the high strength stiffness and low weight can be most useful. In addition to wing sections, elevator and rudder sections are also fabricated in this way. High relative costs have so far precluded widespread use of automatic machine lay-up in the automobile industry.

**Table 8.1**
Tape dimensions

| Parameter | Typical range |
|---|---|
| Thickness, mm (in. $\times 10^{-3}$) | 0.08–0.25 (3–10) |
| Resin content, % | 28–45 nominal $\pm 2$ |
| Dry fiber weight per unit area, g/m$^2$ (oz/ft$^2$) | 30–300 (0.10–1.0) |
| Width, mm (in.) | 25–1525 (1–60) |
| Package size, kg (lb) | 4.5–225 (10–500) |

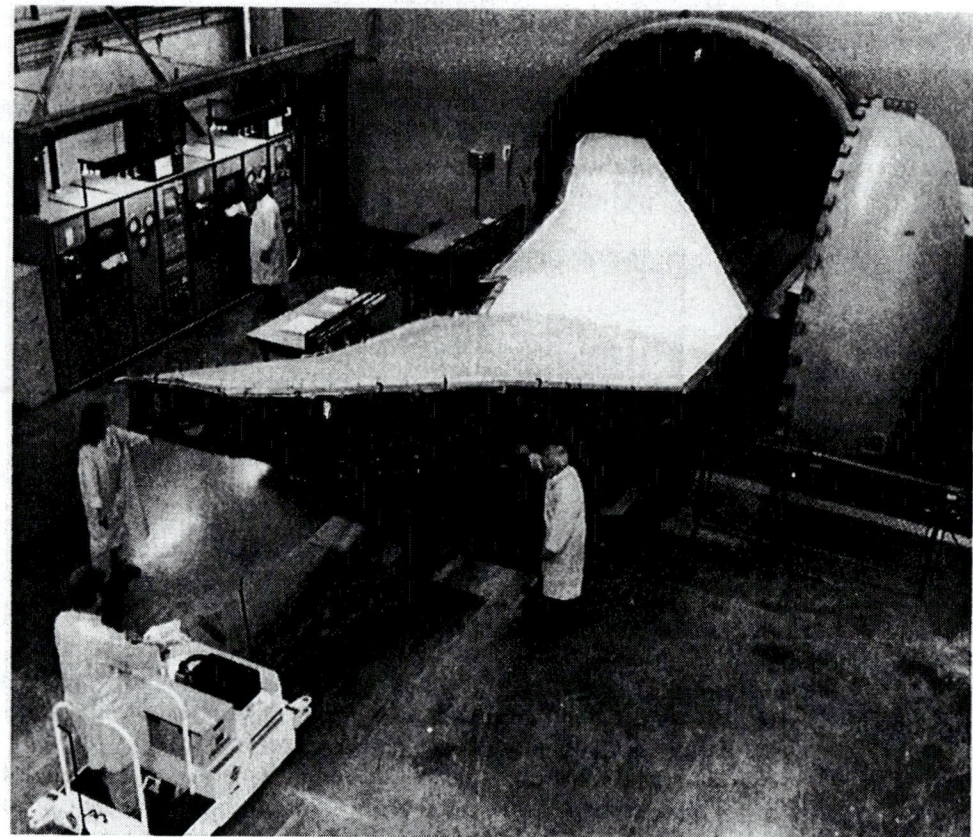

**Figure 8.3**
Carbon fiber–epoxy laminate being placed into autoclave for curing
(From W. F. Smith, *Principles of Materials Science and Engineering*, 2nd ed., McGraw-Hill
Publishing Company, 1990.)

## 8.1.2   *Compression Molding*

**Sheet molding** is the *compression molding* process used mainly in producing glass
fiber–reinforced polyester resin parts. The starting material, which we call *sheet
molding compound (SMC)*, is a thermosetting resin filled with chopped or continuous strands of glass fiber. An SMC processing machine, such as the one shown
schematically in Figure 8.4, produces molding compound in sheet form. The size of
the machine is designated by the width of the sheet produced, the most common
being 4 ft wide.

The process starts in the paste reservoir below the chopper, which meters a
specified amount of resin filler paste onto a plastic carrier film. The paste consists of
several ingredients that are varied to control matrix properties. The carrier film

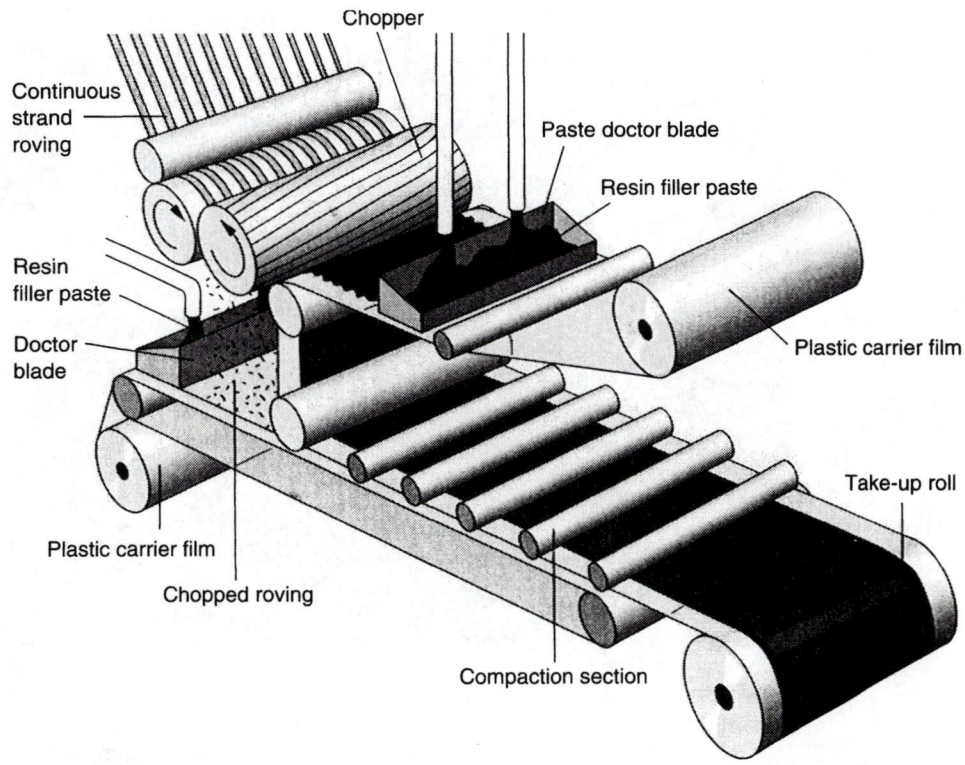

**Figure 8.4**
Sheet molding compound processing machine
(From *Engineered Materials Handbook*, Vol. 1: *Composites*, ASM International, 1987.)

passes under a chopper, which cuts glass fiber roving into 1-in. lengths. After the glass falls to the resin bed, another carrier film with another layer of paste is added on top, sandwiching the glass between the two layers.

When the paste is first mixed and put in the SMC machine, it has the consistency of pancake batter. After partial curing, when the thickening agents have reacted, the material resembles caulking compound, and the carrier film is removed. The SMC material is then cut into charges that are placed into the matched metal die molds of a hydraulic press. Application of heat (150°C) with pressure causes flow of the SMC to all sections of the mold and activates curing (cross-linking) of the material. The part is then removed from the mold.

This process is one of the newer closed mold processes used to produce parts for the automobile industry. It provides good resin control and good mechanical strength properties while producing high-volume, large-size uniform products. Figure 8.5 shows the front hood of an automobile made by this process. The advantages of the SMC process over hand lay-up include more efficient high-volume production, improved surface quality, and uniformity of product.

**Figure 8.5**
Automobile hood panel made by pressing SMC
(From W. F. Smith, *Principles of Materials Science and Engineering*, 2nd ed., McGraw-Hill
Publishing Company, 1990.)

# 8.1.3  Filament Winding

**Filament winding** was developed to provide high-speed, precise lay-down of continuous reinforcement in a specific pattern. Filament winding is a process in which continuous resin-impregnated strands of fiber called tows or rovings are wound over a
rotating mandrel as shown in Figure 8.6.

We can use a mandrel that is cylindrical or any other shape that does not have
any reentrant curvature and we can wrap the reinforcement either in adjacent or
repeating bands that are stepped the width of the band. The machine must have the
capacity to vary winding tension, winding angle, and resin content in each layer of
reinforcement; this ensures the desired thickness and resin content of the composite
in the desired direction of strength. A wound part is cured at room temperature or
in an oven, then the molded part is stripped from the mandrel.

The most important advantage of filament winding is lower cost compared to
prepreg for most composites. The high degree of fiber orientation and high fiber
loading with this method produce very high tensile strengths in hollow cylinders,
leading to such applications as rocket motor cases, pressure vessels, and chemical
storage tanks.

**Figure 8.6**
Filament-winding process for producing fiber-reinforced-plastic composite materials. The fibers are first impregnated with plastic resin and then wound around a rotating mandrel (drum). The carriage containing the resin-impregnated fibers traverses during the winding, laying down the impregnated fibers.
(From W. F. Smith, *Principles of Materials Science and Engineering*, 2nd ed., McGraw-Hill Publishing Company, 1990.)

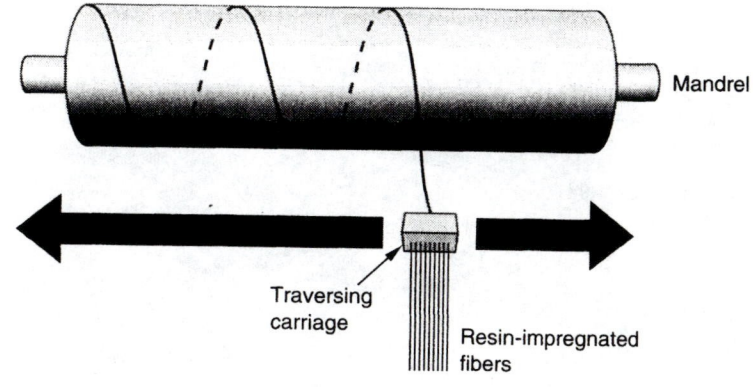

Mandrel

Traversing carriage

Resin-impregnated fibers

## 8.1.4 Resin Transfer Molding

Injection molding, described for polymers in Chapter 3, has been modified to fabricate composites ranging from simple, low-performance parts to complex, high-performance applications of all sizes. For **resin transfer molding** of composites, we place preshaped fiber reinforcement into a tool cavity. After closing, we inject liquid resin to impregnate the reinforcement. A schematic setup of transfer molding is shown in Figure 8.7 for the fabrication of a glass fiber component. The fiber content can be in the range of 35 to 60 weight percent. We use steel tooling for the moderate molding pressures (640–345 kPa, or 100–500 psi) and molding is done at elevated temperature to minimize cure time. Cycle times vary from one minute for small components to 8–12 minutes or longer for large complex structures.

The relationship of resin transfer molding to other processes is shown in Figure 8.8. As the degree of complexity and size of the components increase, labor-intensive processes, such as hand lay-up, are appropriate for a relatively small number of components. Cost limits the use of this technique for production of large numbers of parts. For high-production manufacturing, we are more likely to choose compression molding and thermoplastic forming processes. The equipment required to produce parts with these technologies is expensive, however, so a great many parts must be made before they are economically justified.

The center and upper right-hand section of Figure 8.8 represent the areas where the maximum potential exists for producing economical high-volume composite structures. This area represents high-performance, large, highly integrated structures produced in medium to high volumes. Automotive molders throughout the United States currently produce composite components such as hoods, deck lids, doors, and grill panels by the resin transfer process.

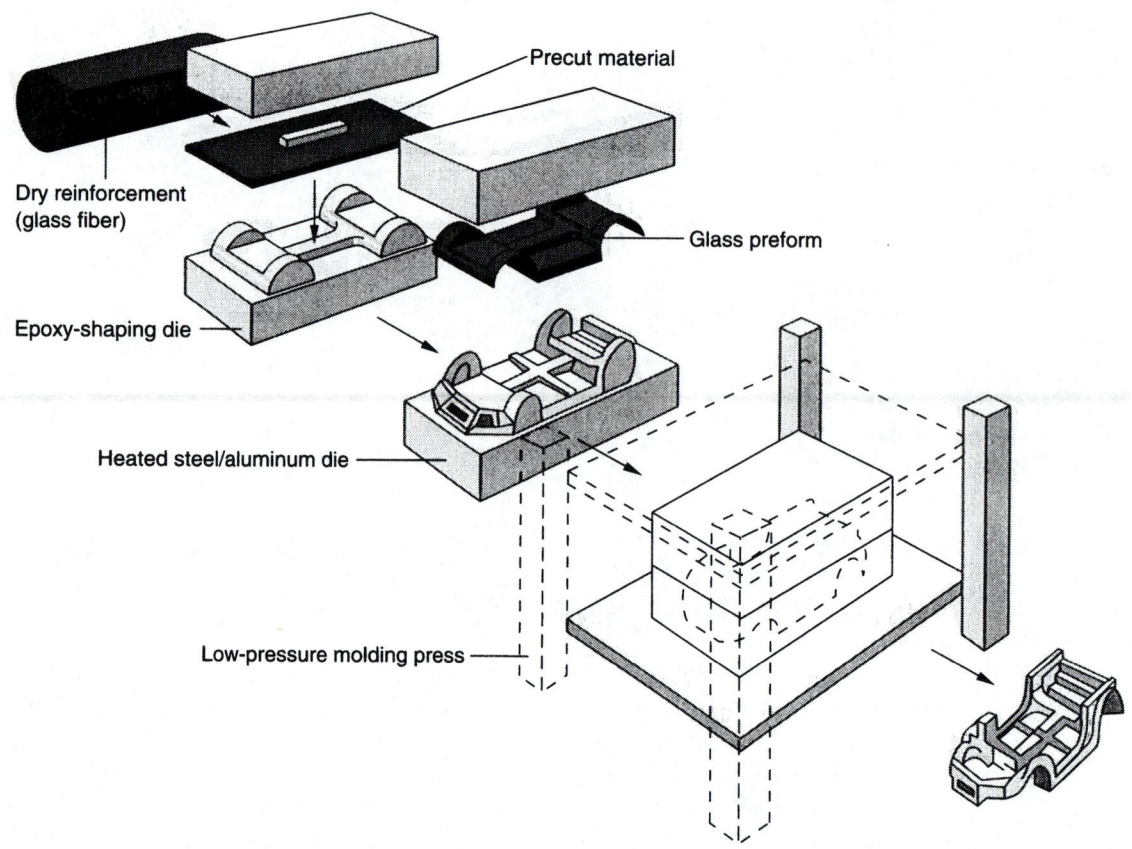

**Figure 8.7**
Resin transfer molding of composites
(From *Engineered Materials Handbook*, Vol. 1: *Composites*, ASM International, 1987.)

## 8.1.5 Carbon Fiber-Reinforced Carbon-Matrix Composites

These materials have many of the desirable high-temperature properties of conventional carbons and graphite, including high strength, high elastic modulus, and low creep. However, they also have good fracture toughness coupled with low expansion coefficients, so they can survive severe thermal and mechanical shock conditions. **Carbon-matrix carbon-fiber composites** were developed to withstand the severe requirements of reentry vehicles in the space program. They were originally developed in the early 1960s as two-directional composites based on heat-treated carbon fabric–phenolic resin laminates. The availability of high-strength, high-modulus car-

**Figure 8.8**
Chart of process development for automotive composite parts: steep slope of the hand lay-up process reflects the high-cost and low-volume characteristics of a labor-intensive process
(From *Engineered Materials Handbook*, Vol. 1: *Composites*, ASM International, 1987.)

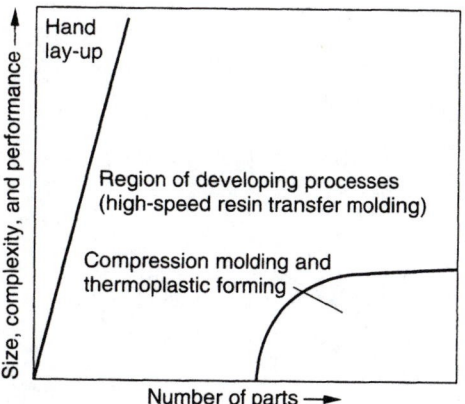

bon fibers in the mid to late 1960s combined with the capability to weave multidirectional carbon fiber preforms led to the development of these composites.

An important advantage of a controlled multidirectional distribution of fibers is the freedom to orient them to accommodate the design loads of the final structural component. Multidirectional fabrication technology provides the means to produce tailored composites. The simplest fiber preform is based on a three-directional construction, which is normally used to weave rectangular block shapes, as shown in Figure 8.9. Preforms consist of multiple yarn bundles located on orthogonal axes.

Originally, the **three-directional orthogonal reinforcement** weaving operations were carried out manually to fabricate the preforms. Although many of these operations are now automated, details regarding equipment and procedures are often proprietary. The techniques used to manufacture these preforms include weaving dry yarns with a three-directional loom for block shapes and modified filament winding for cylindrical shapes, as shown in Figure 8.10.

Densification of the fiber preforms involves infiltration by a precursor polymer that is then decomposed to yield a carbon matrix. Two general categories of matrix precursors used for carbon-carbon densification are thermosetting resins (such as phenolics) and pitches based on coal tar and petroleum. The thermosetting resins polymerize to form cross-linked infusible solids. As a result of pyrolysis, these resins then form carbon. The carbon yield at 800°C is about 50–60 weight percent.

Coal tar and petroleum pitches are mixtures of polynuclear aromatic hydrocarbons that undergo various changes upon heating, yielding about 50 weight percent carbon after pyrolysis at 600°C and atmospheric pressure. Figure 8.11 shows a flow diagram for a typical densification process.

Originally developed as reentry vehicle nosecone materials and rocket motor components, carbon-carbon composites are also used for aircraft brakes and form the nosecone and leading edges of the space shuttle vehicle. Because these materials are biocompatible, they can be tailored to be structurally compatible with bone for the fixation of fractures and for hip joint replacements.

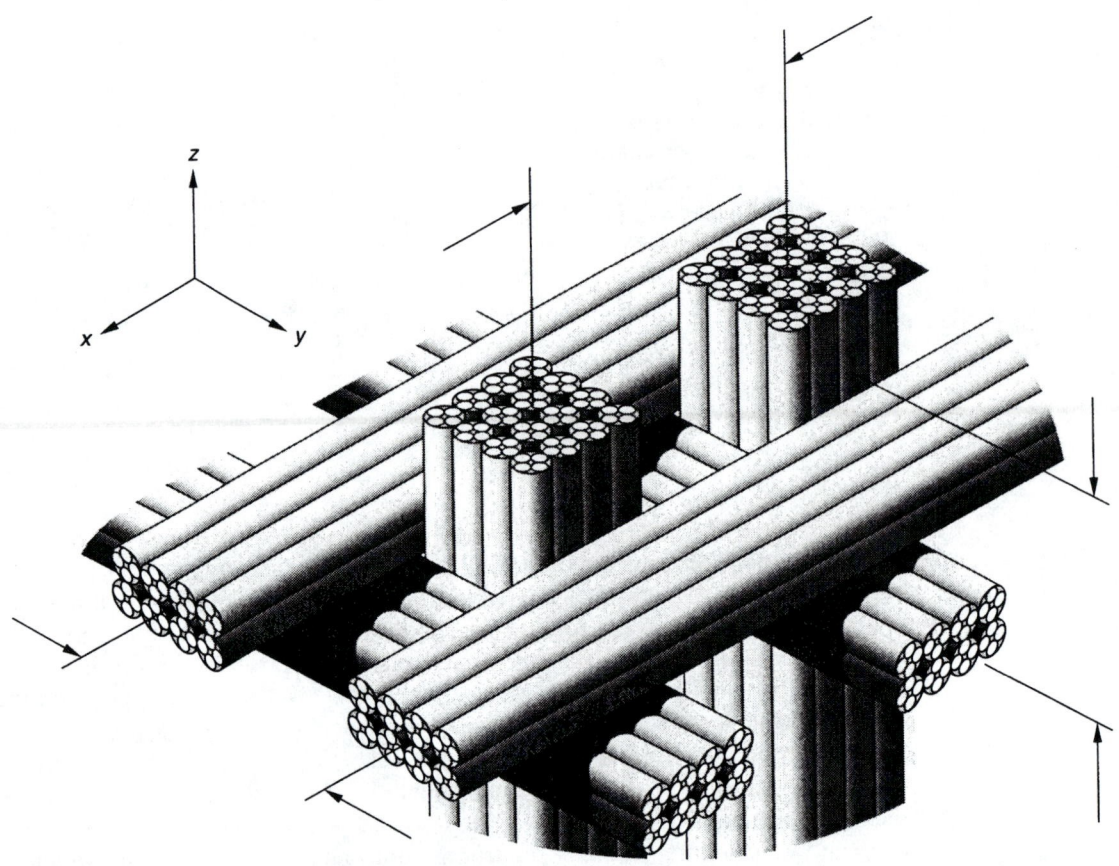

**Figure 8.9**
Three-directional orthogonal preform construction
(From *Engineered Materials Handbook*, Vol. 1: *Composites*, ASM International, 1987.)

# 8.2   Processing of Metal Matrix Composites

Two types of **metal matrix composites** are continuous fiber and discontinuous fiber or whisker reinforced. These composites have been developed largely for the aerospace industry but some are being used in other applications as well.

## 8.2.1   Continuous Fiber-Reinforced Metal-Matrix Composites

Continuous filaments provide the greatest potential improvement in mechanical properties for metal matrix composites if significant fiber damage does not occur. Currently, high-pressure diffusion bonding is the only commercial fabrication tech-

**Figure 8.10**
Automated three-directional preform fabrication equipment
(From *Engineered Materials Handbook*, Vol. 1: *Composites*, ASM International, 1987.)

nique used in the manufacture of boron fiber–reinforced metal composites. Other processes such as casting have been examined but the degradation of mechanical properties of boron, particularly in an aluminum matrix at about 525°C, limits the feasibility of this method without the use of expensive surface coatings. For the diffusion-bonding process, a preform consisting of boron filaments and metal foil is first prepared by hot-pressing an array of the fibers between metal foil layers. At elevated temperatures, the foils deform around the fibers and bond to the fibers and to each other. These preforms can then be combined to form simple structures. Variations of the process include plasma spraying of metal onto the fiber array to make

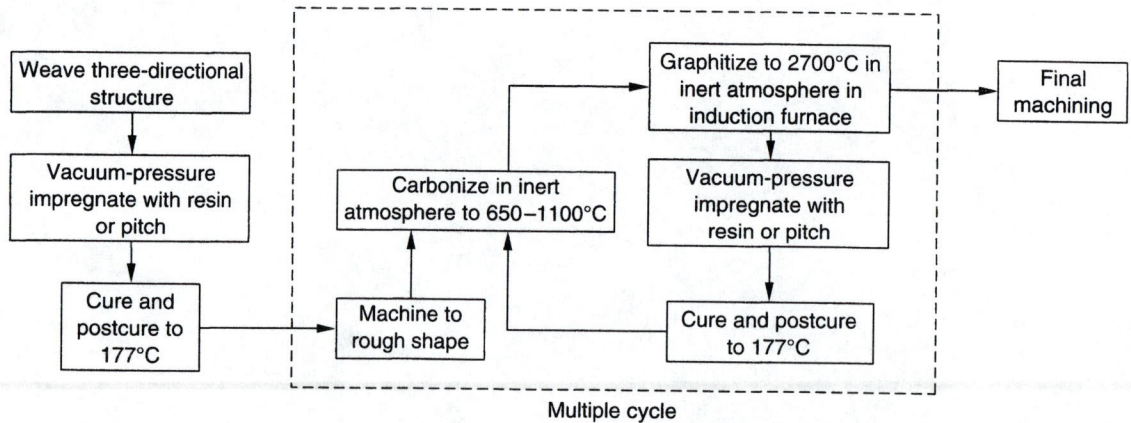

**Figure 8.11**
Typical carbon-carbon process
(From *Engineered Materials Handbook*, Vol. 1: *Composites*, ASM International, 1987.)

the preform and the step-pressing of continuous boron fiber–metal preform tape. In the latter case, discrete segments of fiber-foil sandwiches are sequentially diffusion bonded, producing a preform that is continuous in the fiber direction. These preforms are typically fabricated using 140 micron (0.0056 in.) diameter fiber. Figure 8.12 shows the typical fabrication sequence to produce boron-aluminum composites.

Silicon carbide fiber–reinforced metals can be produced more easily because these fibers readily bond to metals and resist strength degradation better than boron during high-temperature processing.

Intermediate products such as preforms and fabrics used in component fabrication are produced first to simplify the loading of fibers into a mold and to provide correct alignment and spacing of the fibers. The preparation of such "green tape" involves winding fibers onto a rotating drum covered with foil and overspraying with resin. The layer is cut from the drum to provide a flat sheet of prepreg. Prepreg lay-ups are then loaded into a mold in required orientations to fabricate laminates. The laminate processing cycle is controlled so as to remove the resin by volatilization using both heat and vacuum.

Plasma-sprayed aluminum tape is similar to green tape except that the resin binder is replaced with a plasma-sprayed matrix of aluminum. Advantages of this prepreg include absence of contamination from resin residue and faster material processing times due to elimination of the hold time required to remove resin binder. As with the green tape system, lay-ups of the plasma-sprayed preforms are prepared sequentially in the mold and pressed to final shape at low pressure. This process was designed to fabricate shaped silicon carbide fiber–aluminum parts at significantly lower cost than would be possible by a diffusion bonding–solid-state process. Because the silicon carbide fibers can withstand molten aluminum for long periods, the molding temperature can be raised into the liquid plus solid region of the alloy to ensure aluminum flow and consolidation at low pressure, thereby eliminating the need for high-pressure die molding equipment. This hot-molding process is analogous to the autoclave molding of graphite fiber–epoxy composite, in which components are molded in an open-faced tool. The mold in a metal composite, however, is a slip cast ceramic. A composite preform is laid into the mold, heated to near molten aluminum temperature, and pressurized in an autoclave, as illustrated in Figure 8.13.

Diffusion bonding of silicon carbide–titanium is accomplished by hot-pressing using fabric fiber preforms that are stacked together between titanium foils for consolidation. Two methods are being developed by aircraft and engine manufacturers to make complex shapes. One method based on hot isostatic pressing technology uses a steel pressure membrane to consolidate components directly from the fiber-metal preform layer. The other method requires the use of a previously hot pressed silicon carbide fiber–titanium substructure during subsequent forming operations. A typical use of the hot isostatic pressing procedure, for a SiC-titanium engine drive shaft, is illustrated in Figure 8.14. The fiber preform is placed onto a titanium foil that is then spirally wrapped, inserted, and diffusion bonded onto the inner surface of a steel tube using a steel pressure membrane. The steel is thinned down and machined to form the spline attachment at each end.

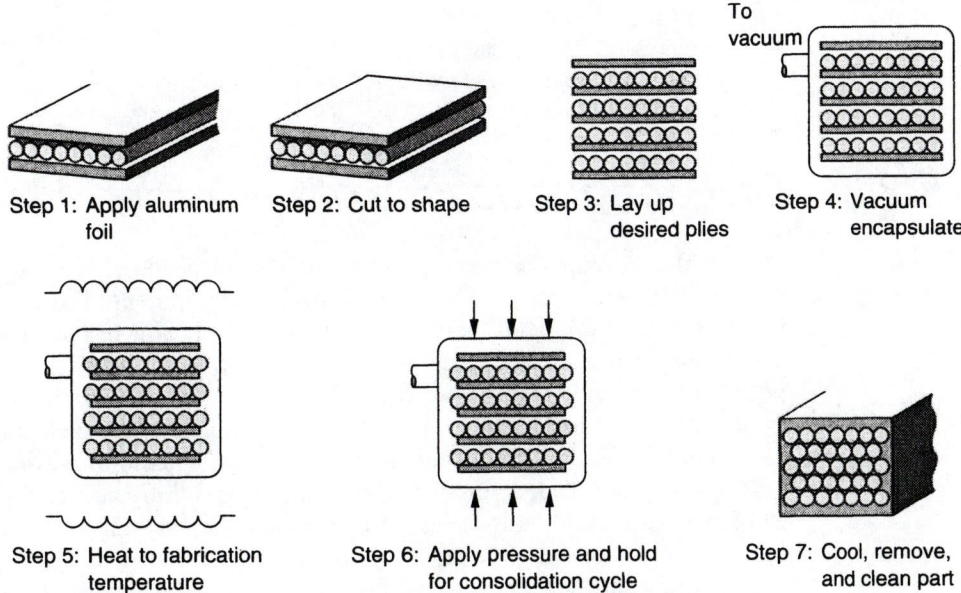

Step 1: Apply aluminum foil

Step 2: Cut to shape

Step 3: Lay up desired plies

Step 4: Vacuum encapsulate

Step 5: Heat to fabrication temperature

Step 6: Apply pressure and hold for consolidation cycle

Step 7: Cool, remove, and clean part

**Figure 8.12**
Typical fabrication process for boron-aluminum composites
(From *Engineered Materials Handbook*, Vol. 1: *Composites*, ASM International, 1987.)

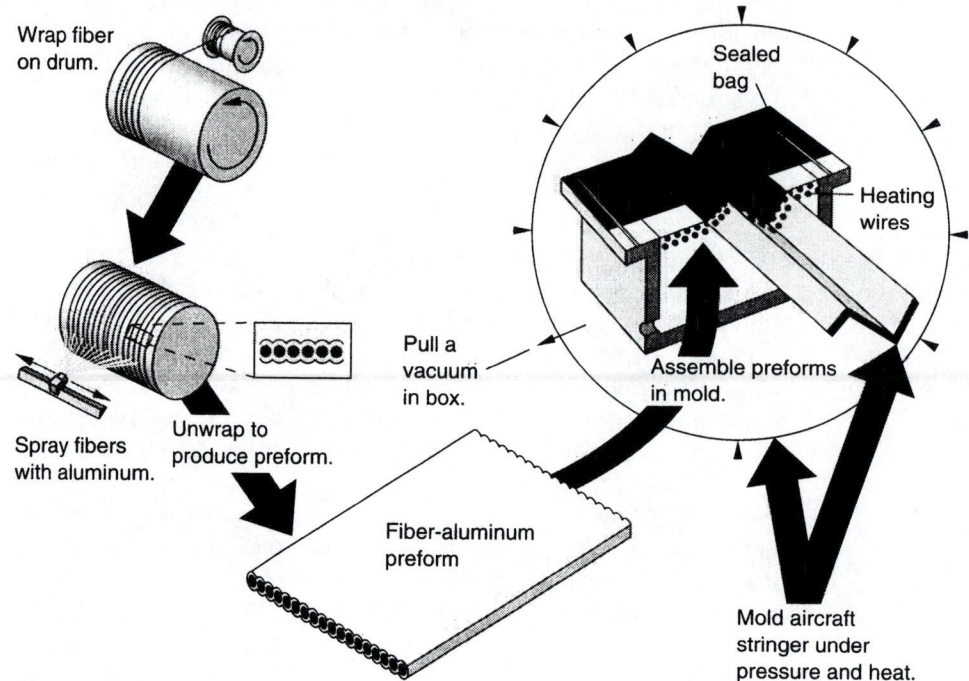

**Figure 8.13**
Hot molding of SiC-Al Z stiffeners
(From *Engineered Materials Handbook*, Vol. 1: *Composites*, ASM International, 1987.)

## 8.2.2  *Discontinuous Fiber-Reinforced Metals*

One of the principal advantages of using discontinuous fibers or whiskers is the opportunity to use conventional metal-forming equipment. The preparation of SiC whisker–reinforced metal-matrix composites in particular has been accomplished by using powder metallurgy techniques as well as melt infiltration. Hot-pressed composites require more effort, but typically have properties superior to those of melt-infiltrated composites. Despite the ease of manufacture, the melt-infiltrated materials usually have a nonhomogeneous distribution of fibers or whiskers in the metal matrix and flaws also occur in the matrix during solidification. Figure 8.15 shows whisker-reinforced metals prepared by vacuum hot pressing to form a dense, pore-free billet that can then be shaped by secondary processing.

In melt infiltration, a fiber or whisker preform is prepared and infiltrated by molten metal. Subsequent solidification produces the fiber-reinforced metal-matrix composite. Preforms are produced either by a process similar to ceramic slip casting or by a pulp-molding process. Whiskers are first combined with various binders to

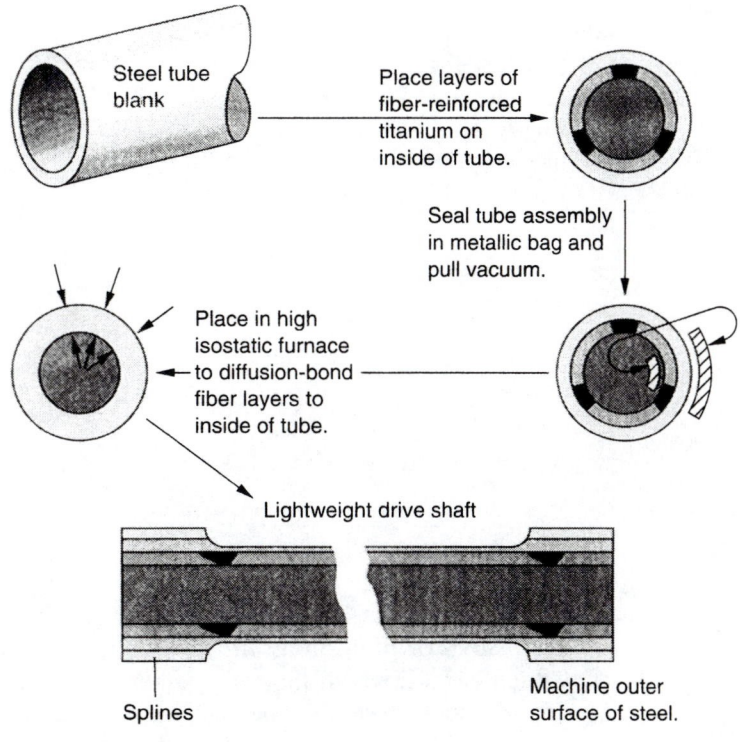

**Figure 8.14**
Hot isostatic pressing of SiC-Ti drive shaft
(From *Engineered Materials Handbook*, Vol. 1: *Composites*, ASM International, 1987.)

Steel tube blank

Place layers of fiber-reinforced titanium on inside of tube.

Seal tube assembly in metallic bag and pull vacuum.

Place in high isostatic furnace to diffusion-bond fiber layers to inside of tube.

Lightweight drive shaft

Splines

Machine outer surface of steel.

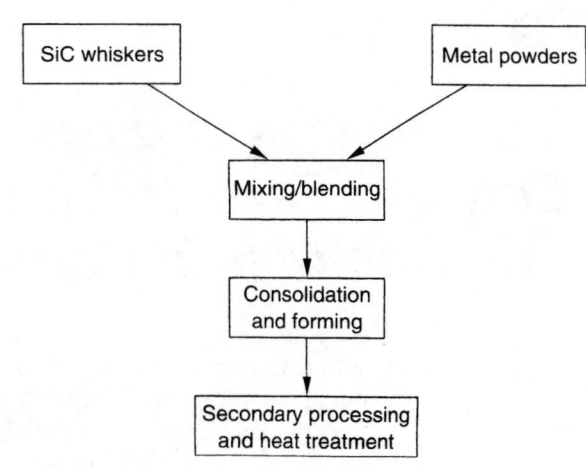

**Figure 8.15**
Powder metallurgy process for SiC whisker–reinforced metal-matrix composites

SiC whiskers

Metal powders

Mixing/blending

Consolidation and forming

Secondary processing and heat treatment

*159*

**Figure 8.16**
Melt infiltration process for
whisker-reinforced metal-
matrix composites

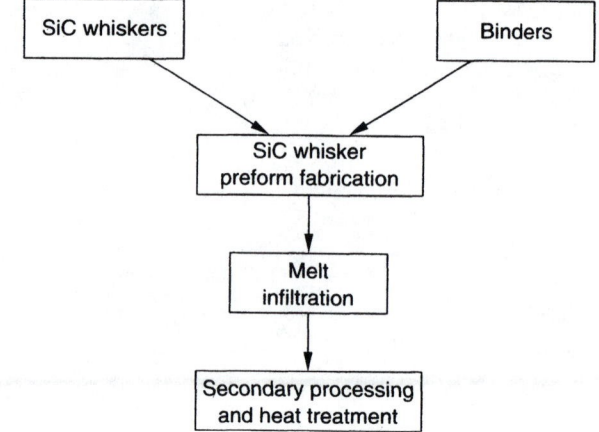

form a slurry that is poured into mold cavities or vacuum molded and allowed to cure. Mold or preform assemblies are then furnace fired to burn out organic constituents and drive off moisture. When the firing is complete, the delicately bonded and interlocked whiskers become a cohesive, porous preform body. These preforms are 20–40% dense and have sufficient strength to be placed in the mold for final processing. Figure 8.16 shows the process flow for the melt infiltration process.

An important feature of SiC whisker–reinforced composites is that they can be reprocessed using conventional metalworking equipment such as extrusion, forging, and rolling facilities. However, these materials require greater control of processing temperature and strain rate than unreinforced metals because of the whiskers present.

# 8.3   Processing of Ceramic Fiber-Reinforced Ceramic-Matrix Composites

In Chapter 7, we saw that both continuous and discontinuous ceramic fiber–reinforced **ceramic-matrix composites** have been developed for specific applications. Continuous fiber composites are first prepared as preforms or as three-directional fabric plies. These preforms have been made from SiC fibers, fused silica fibers, and carbon fibers. Matrix infiltration in three-directional preforms is usually accomplished by either chemical vapor infiltration or by pressure infiltration. Heat treatment converts the polymer to a ceramic matrix phase through decomposition. Polycarbosilane, for example, is a good precursor for obtaining silicon carbide. Matrix buildup by chemical vapor infiltration is accomplished by allowing reactant gases to make contact with a preheated preform, causing deposition of a matrix phase on the fiber preform surfaces. For deposition of silicon carbide at about 1000°C, one reaction is

$$SiCl_4 + CH_4 \rightarrow SiC + 4\,HCl$$

Because of surface pores filling up preferentially, this process often requires several cycles to reach even 10–15% residual pore volume. After each cycle, sample surfaces have to be machined away in order to remove the dense surface layer so that further infiltration can be performed successfully. The process is slow, often requiring days of deposition time to densify a preform of about 40 volume percent fiber to obtain 70–80% of theoretical density.

Two-directional preforms in the form of fabric plies can be surface loaded with matrix phase powder, with each ply sandwiched between two matrix phase layers. These are stacked together in a graphite mold and are compressed, bonded, and sintered by hot pressing. This technique is suitable for the processing of ceramic composite panels containing continuous fibers. Applications for these continuous fiber composites have been almost exclusively associated with the aerospace industry as heat shield components and radar-transmitting windows.

Discontinuous fiber or whisker–reinforced ceramic composites are prepared by particulate processing methods with a key step being the uniform blending and distribution of the short-length fiber or whisker phase in the ceramic powder matrix. This parallels the processing of metal matrix powders with whiskers described earlier in Section 8.2.2. After blending, the mixture of ceramic powder and fiber or whiskers is sintered in graphite molds by hot pressing. One composition based on SiC whiskers in an $Al_2O_3$ matrix has been commercially successful as a cutting tool for hardened steels.

# Summary

The processing of fibrous composite materials is achieved by methods that are largely based on fiber length and configuration as well as the matrix type. Polymeric matrix composites are most common because of reasonable cost, relative ease of manufacture, and large weight savings. Thermosetting resins are the matrix of choice because of bonding and forming characteristics. High-temperature limitations of the polymer-based composites have spurred the development of advanced metal matrix composites and ceramic matrix composites.

# Terms to Remember

carbon-matrix carbon-fiber composites

ceramic-matrix composites

collimated

compression molding

fiber reinforcement

filament winding

lay-up process

metal matrix composites

plastic matrix composites

prepregging

resin transfer molding

sheet molding

tape

three-directional orthogonal reinforcement

tooling

# Problems

1.  What are the important applications of plastic matrix composites fabricated by the lay-up process?
2.  Why is the use of tapes important to the lay-up process?
3.  What are the primary materials used in sheet molding?
4.  Why was the filament-winding process developed?
5.  Describe resin transfer molding. What is the justification for this process?
6.  What is the justification for carbon-carbon composites? Outline their fabrication process.
7.  Describe two categories of metal matrix composites and their processing feature.
8.  Describe a process for fabricating a ceramic matrix composite.

# 9

# *Concrete*

Concrete, along with steel and wood, is one of the most common construction materials. **Concrete** is used worldwide for roadways, dams, buildings, barriers, architectural surfaces, and other applications too numerous to list. Reasons for concrete's popularity include its excellent strength in compression, inexpensive cost, ease of shaping, and excellent water resistance. It is both a ceramic, because it is a brittle inorganic nonmetallic material, and a composite, because it comprises aggregates bonded together by hydraulic cement, materials that are readily available anywhere in the world. Concrete is easy to shape because ingredients are mixed with water, forming a plastic semifluid that can be readily formed before curing to a firm, strong solid. In this chapter, we will examine the types of concrete, each of the components in this remarkable material, and control of properties for specific applications.

## 9.1 Types of Concrete

As a composite material, the amounts and distribution of each component can alter the characteristics of concrete. For example, the terms *normal*, *lightweight*, and *heavyweight* are used to describe concrete, but the differences, of course, derive from the density of the aggregates used. It is more useful to also look at concrete in terms of its **compressive strength**, the property of interest for most applications. Low-strength concrete has less than 3000 psi compressive strength, moderate-strength concrete (most normal weight used for structural purposes) has 3000–6000 psi compressive strength, and high-strength concrete has greater than 6000 psi compressive strength.

There are many other special types of concrete and their names appropriately describe their special uses. For example, there are steel rods in reinforced concrete and prestressed concrete contains reinforcing strands that maintain cured concrete in compression.

# 9.2  Concrete Ingredients

We learned in studying composite materials that strength is derived from transfer of stress from a weak matrix to a stronger component. In concrete, aggregates are the stronger components; they make up 60–80% of the volume of concrete and influence plasticity, cured strength, **durability**, and dimensional stability, as well as cost of the concrete. Cement is the matrix that bonds to these aggregates during curing, thus transferring stress to them during use.

## 9.2.1  Aggregates

**Aggregates**, for the most part, are natural materials such as sand, gravel, and crushed rock. Those larger than 4.75 mm ($^3/_{16}$ in.) are termed *coarse* and *fine* aggregates are smaller than this measurement. Sand and gravel are important natural unconsolidated sediments that need only washing and classification for use in concrete. Coarser aggregates, however, are usually crushed rock, whose properties depend to a large extent on impurities, on **porosity**, and on friability as well as on hardness or strength. Most normal-weight concrete aggregates are crushed carbonate rocks such as dolomite and limestone. Where available, crushed igneous rocks such as granite are preferable because of their strength, low porosity, and crushed shape. Table 9.1 lists some concrete properties that are dependent on aggregate properties.

When lightweight concrete is to be made, aggregates selected are very porous inorganic materials, such as expanded shale or pumice (a natural volcanic rock), or synthetic materials, such as blast furnace slag. Heavyweight concrete is made from aggregates derived from steel scrap, from rocks containing barium minerals, or from ores such as iron ore.

**Table 9.1**
Influence of aggregate properties on concrete properties

| Concrete property | Aggregate property |
|---|---|
| Strength | Strength, cleanliness, surface texture, size, shape |
| Durability | |
|   Abrasion resistance | Hardness |
|   Freeze-thaw resistance | Porosity, permeability, tensile strength |
| Thermal resistance | Thermal expansion |
| Slipperiness | Surface texture, size, shape |

# 9.2.2 Cement

Aggregates of various sizes and distributions are easy to recognize if we examine polished concrete, shown in Figure 9.1, but the lighter matrix of cement is not readily distinguishable. Cement bonds aggregates together to form concrete. It is also responsible for one of the most important properties of concrete, its resistance to water. Cements that resist water when cured are called **hydraulic cements** to distinguish them from other cementitious materials like **gypsum** that cure but are unstable in water. Portland cement is the main hydraulic cement used in concrete production, but there are natural minerals or pozzolans that are also hydraulic cements. **Portland cement** is manufactured by high-temperature chemical reaction of limestone and clay as main materials to form clinkers that are then ground into fine powder. The cement consists of various oxides of calcium, silicon, aluminum, and iron, but the amounts of each are variable to control curing rate or heat generated during curing. For practical purposes, consider portland cement to be made up of the following four principal compounds (listed with their chemical formulas and common abbreviations):

| | | |
|---|---|---|
| tricalcium silicate | $3CaO \cdot SiO_2$ | $C_3S$ |
| dicalcium silicate | $2CaO \cdot SiO_2$ | $C_2S$ |
| tricalcium aluminate | $3CaO \cdot Al_2O_3$ | $C_3A$ |
| tetracalcium aluminoferrite | $4CaO \cdot Al_2O_3 \cdot Fe_2O_3$ | $C_4AF$ |

The types of portland cement and their approximate compound compositions are given in Table 9.2.

**Curing**, or hardening, of the cement is a chemical **hydration** that incorporates water into the structure (the product of the hydration is abbreviated as H). This is a complicated process that produces the solid but inhomogeneous cement binder of the concrete composite. Each of the ingredients of the portland cement hydrates at a different rate, thus affecting the strength as curing time increases. For example, **$C_3A$** sets very quickly and can even interfere with shaping the concrete, but the addi-

**Figure 9.1**
Low magnification photomicro-
graph of polished concrete
section
(From S. H. Kosmatka and
W. C. Panarese, *Design and
Control of Concrete Mixtures,*
13th ed., Portland Cement
Association, 1988.)

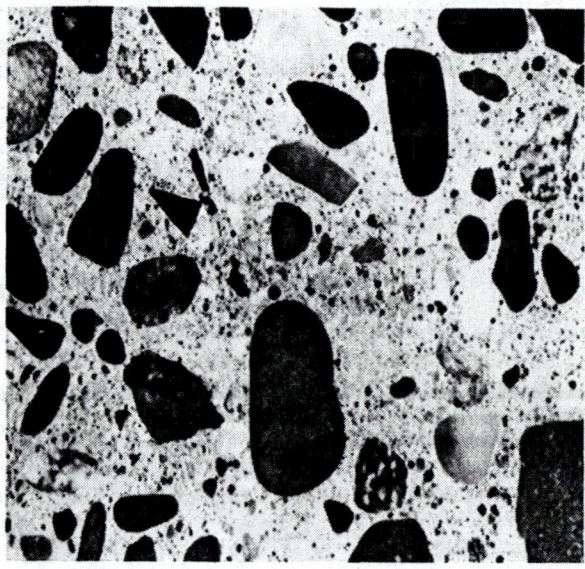

tion of gypsum slows this process, thus promoting formability. Gypsum, however, also provides sulfates that dissolve and take part in hydration of aluminates. High sulfate concentration leads to the formation of **ettringite**, which are needlelike mineral crystals. As sulfate is depleted, however, ettringite becomes unstable and forms **monosulfate**.

$C_3S$ and $C_2S$ hydrate at slower rates and both form **calcium silicate hydrates**, abbreviated **C-S-H**, which are poorly defined tobermorite crystals with a very high surface area-to-volume ratio, and **calcium hydroxide** (CH), which is made up of large platelike crystals. C-S-H makes up more than half of the solids formed during curing and is largely responsible for strength. However, $C_3S$ forms less C-S-H and more CH than $C_2S$, so cements with higher $C_2S$ are stronger. CH makes up 20–25% of the solids formed by the hydration, but these do not contribute significantly to strength because of their low surface area-to-volume ratio.

**Table 9.2**
Principal types of portland cement

| Type | Utility | Compounds (approx. wt %) | | | |
|------|---------|------|------|------|------|
| | | $C_3S$ | $C_2S$ | $C_3A$ | $C_4AF$ |
| Type I | Normal | 55 | 19 | 9 | 7 |
| Type II | Sulfate resistance | 51 | 24 | 6 | 11 |
| Type III | High early strength | 56 | 19 | 9 | 7 |
| Type IV | Low heat of hydration | 28 | 49 | 4 | 12 |
| Type V | Severe sulfate resistance | 38 | 43 | 4 | 9 |

The products of hydration do not tell us anything about the rate at which hydration takes place or the heat generated during the reaction. C₃S, C₃A, and **C₄AF** all hydrate more rapidly than C₂S and, of course, form more heat during the early curing times. This is beneficial for cold-weather concreting, but can pose problems for massive structures. This is true because heat cannot readily be dissipated and resulting nonuniform cooling can produce stresses that lead to cracking.

The development of hydration products for a Type I ordinary cement paste and the change in the characteristics as this development takes place are summarized in Figure 9.2. Similar data could be developed for other types of portland cement now that we know how hydration proceeds.

**Figure 9.2**
Effect of hydration time for Type I portland cement (a) on formation of hydration products and (b) on characteristics of the products (From P. K. Mehta and P. J. M. Montiero, *Concrete: Structure, Properties, and Materials,* 2nd ed., Prentice Hall, 1993.)

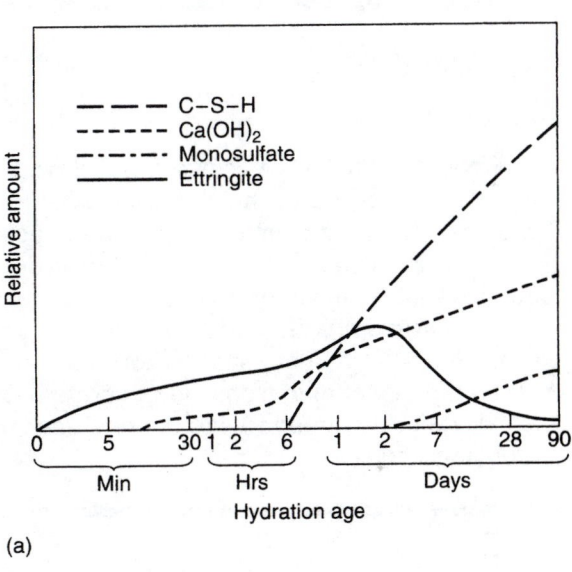

(a)

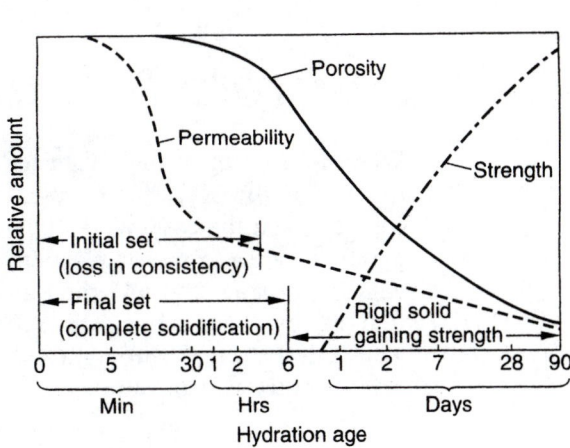

(b)

## 9.2.3   *Water*

Almost any natural water that is drinkable and has no profound taste or smell can be used as mixing water for concrete. Water suitable for making concrete, however, may not be safe for drinking. Some impurities do retard the hydration process and should be avoided, for example, sugars or ethylene glycol, organic matter, and soluble salts of zinc, lead, copper, and other metals. Seawater is used in some instances because the chloride promotes early strength, but the sulfates tend to slow hardening because they prolong the ettringite crystallization. Potential accelerated corrosion must be considered to avoid the problems encountered in Case Study 9.1.

---

### Case Study 9.1

#### Corrosion Failure

In the construction of an above-ground water storage tank in rural Maine, radiant heating pipes were installed in the concrete slab to prevent freezing during the winter months. Plans specified mild steel pipes, but no specifications for concrete mixing water were considered. Located near a salt marsh, concrete construction workers took advantage of the readily available saltwater that would not freeze as hydration took place.

Construction was completed in early spring, but the radiant heating never did take place. When the heating season approached and hot water pipes were laid, it was found that the pipe extending from the concrete broke off because of extensive corrosion. Failure analysis quickly pointed to salt content and moisture as the cause of the unanticipated corrosion conditions!

---

## 9.3   *Concrete Mixtures* _____

When we formulate the recipe for specific concrete mixtures, we have to consider the **workability** of fresh concrete, the durability, strength, and appearance of cured concrete, and the cost. Workability is necessary for placing, compacting, and finishing the concrete; it is complicated, depending on such factors as water content, the aggregate shape, and absorbed moisture content. The sand/aggregate ratio comes into play because sand improves the finishing characteristics whereas coarser aggregates affect strength. Strength of the final product depends on the cement content *and* the **water/cement ratio**.

From an economic standpoint, the most expensive component is the portland cement. This can cost much more than the aggregate and so must be a prime consideration. The first property to consider is workability, which includes both consistency and cohesiveness, or plasticity. Consistency refers to the ability of fresh concrete to flow, usually measured by the slump test whereby a 12-in. high cone is filled with concrete, compacted, inverted, and then the cone is removed. The concrete slumps and the difference between its final height and the original height of 12 in. is referred to as the **slump**. Slump is higher, of course, for wetter mixtures, but segregation of the water in the mix can also occur and must be avoided. Minimum slumps are 1 in. and values typically do not exceed 4 in. For a given slump, we can alter the amount of water in the mix and alter gradation of the aggregate by selecting aggregate particles based upon their surface roughness. Cohesiveness, or plasticity, is usually determined by how easy the concrete mix can be troweled. We usually can improve cohesiveness by increasing the amount of fines in the aggregates and by increasing the amount of cement.

In practice, most concrete is ordered from a batch plant by phone with specifications given for the slump, maximum aggregate size, and 28-day compressive strength. We should have some ideas of the methods used to determine the batch weights for mixing concrete with the highest slump consistent with placement and compaction without segregation. In order to do this, we must select the aggregate, using a blend in which the maximum size is the largest economically available but no more than one-fifth the smallest dimension of the structure. The amount of water depends on both the desired slump and the maximum aggregate size, which can be obtained from tables. For example, a 3-in. to 4-in. slump with ³/₄-in. aggregate size requires about 340 lb of water per cubic yard of concrete. Once we know the water content, we can determine the amount of cement from the required strength and from tables compiled by the American Concrete Institute.

The proper blend of coarse and fine aggregate depends on both the coarse aggregate size and the degree of fineness of the fine aggregate. Experience has shown that the finer the sand and the larger the coarse aggregate, the higher the coarse aggregate fraction for suitable workability. For example, for ³/₄-in. aggregate size and fine sand, 66% of the total dry concrete volume should be coarse aggregate, but the value decreases to 60% for coarser grades of the fine aggregate. Tables again can be used to estimate the amount of the coarse aggregate, knowing the fineness of the fine aggregate.

Now all that remains is to figure the amount of fine aggregate, which we can do by difference, since we now know the weight of water, cement, and coarse aggregate. If we do not know from experience the density of fresh concrete, we can estimate the total concrete from tables calculated for concrete of medium richness as prescribed by the American Concrete Institute. When all calculations are complete, trial batches are made to fine-tune the recipe until the specified characteristics are achieved.

### Sample Problem 9.1

Using the values given in Table 9.3, determine the batch weights for concrete with the following properties:

Slump is 3 in.

Maximum aggregate size is $3/4$ in. (density, 90 lb/ft³)

28-day compressive strength is 3500 psi

### Solution

a. Slump is given as 3 in.

b. Aggregate size is given as $3/4$ in.

c. From Table 9.3, water content is 340 lb/yd³.

d. In order to calculate the water/cement ratio from the specified compressive strength, standard deviation must be considered. When data are not available, the American Concrete Institute recommends adding 1200 psi for specified strengths of 3000–5000 psi. For 4700 psi, the water/cement ratio is 0.51.

$$\frac{340}{0.51} = 667 \text{ lb cement}$$

e. Table 9.3 gives the volume of the coarse aggregate as 0.62.

$$0.62 \times 27 \text{ ft}^3/\text{yd}^3 \times 90 \text{ lb/ft}^3 = 1507 \text{ lb/yd}^3$$

f. From Table 9.3, the first estimate of fresh concrete weight is 3960 lb/yd³.

$$\text{Sand} = 3960 - (340 + 667 + 1507) = 1446 \text{ lb/yd}^3$$

# 9.4 The Strength of Concrete

We must remember that concrete is a composite material. Its strength is derived from transfer of stress from the weak phase to the stronger one and the effectiveness of the bonding. We know that strength of concrete is affected by the water/cement ratio and by the degree of hydration, illustrated in Figure 9.3, but it is also affected by the volume of voids produced during hydration. Voids result from the shrinkage that takes place during hydration and can occur intentionally, as we shall see later in the discussion of freeze-thaw resistance of concrete.

Porosity weakens the cement paste, but what happens to the ability to transfer strength to the stronger aggregates? This structural relationship can best be under-

**Table 9.3**

Relationships for determination of concrete recipes

| | Approximate mixing water and air content requirements for different slumps and nominal maximum sizes of aggregates | | | | | | | |
|---|---|---|---|---|---|---|---|---|
| | Water, lb/yd³ of concrete for indicated nominal maximum sizes of aggregate | | | | | | | |
| Slump (in.) | ³/₈ in. | ¹/₂ in. | ³/₄ in. | 1 in. | 1¹/₂ in. | 2 in. | 3 in. | 6 in. |
| 1 to 2 | 350 | 335 | 315 | 300 | 275 | 260 | 220 | 190 |
| 3 to 4 | 385 | 365 | 340 | 325 | 300 | 285 | 245 | 210 |
| 6 to 7 | 410 | 385 | 360 | 340 | 315 | 300 | 270 | — |

**Relationships between water-cement ratio and compressive strength of concrete**

| Compressive strength at 28 days (psi) | Water-cement ratio, by weight |
|---|---|
| 6000 | 0.41 |
| 5000 | 0.48 |
| 4000 | 0.57 |
| 3000 | 0.68 |
| 2000 | 0.82 |

**Volume of coarse aggregate per unit of volume of concrete**

| Maximum size of aggregate (in.) | Volume of dry-rodded coarse aggregate per unit volume of concrete for different fineness moduli of sand | | | |
|---|---|---|---|---|
| | 2.40 | 2.60 | 2.80 | 3.00 |
| ³/₈ | 0.50 | 0.48 | 0.46 | 0.44 |
| ¹/₂ | 0.59 | 0.57 | 0.55 | 0.53 |
| ³/₄ | 0.66 | 0.64 | 0.62 | 0.60 |
| 1 | 0.71 | 0.69 | 0.67 | 0.65 |
| 1¹/₂ | 0.75 | 0.73 | 0.71 | 0.69 |
| 2 | 0.78 | 0.76 | 0.74 | 0.72 |
| 3 | 0.82 | 0.80 | 0.78 | 0.76 |
| 6 | 0.87 | 0.85 | 0.83 | 0.81 |

**First estimate of weight of fresh concrete**

| Maximum size of aggregate (in.) | First estimate of concrete weight (lb/yd³) |
|---|---|
| ³/₈ | 3840 |
| ¹/₂ | 3890 |
| ³/₄ | 3960 |
| 1 | 4010 |
| 1¹/₂ | 4070 |
| 2 | 4120 |
| 3 | 4160 |
| 6 | 4230 |

(Courtesy of the American Concrete Institute.)

**Figure 9.3**

Influence of the water-to-cement ratio and curing time on concrete strength (From S. H. Kosmatka and W. C. Panarese, *Design and Control of Concrete Mixtures,* 13th ed., Portland Cement Association, 1988.)

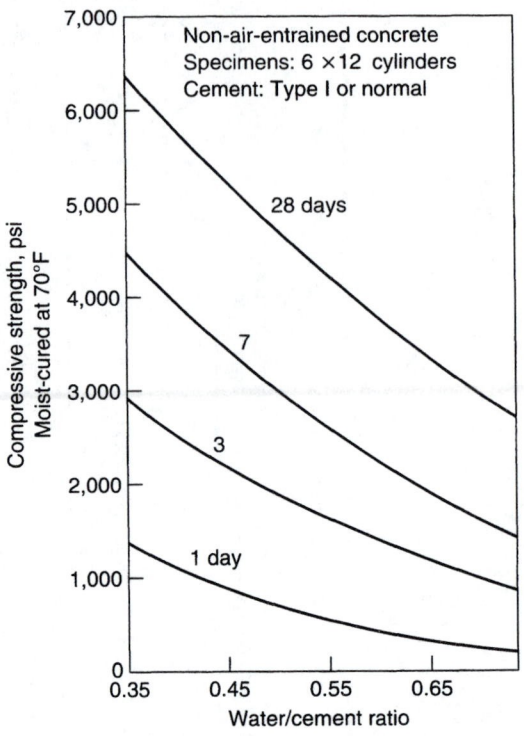

stood by considering the **transition zone** (where bonding takes place) as a third phase, one in which the characteristics change with time, humidity, and temperature. In freshly mixed concrete, water forms a film on the aggregate surfaces, giving a higher water/cement ratio. This high water/cement ratio leads to formation of ettringite and calcium hydroxide and to higher porosity than in the bulk cement paste. The large calcium hydroxide crystals have poorer adhesion characteristics and crack easily; thus the transition zone strength is poor as hydration begins. At later stages of hydration, ettringite dissolves and C-S-H fills in the transition zone, helping to improve its strength. Strength, however, always remains lower in the transition zone than in the bulk cement paste. A schematic diagram of the transition zone appears in Figure 9.4 along with an SEM micrograph of ettringite needles in the fracture surface of a 6-day compression test sample. For 28-day tests, fracture surfaces show no evidence of ettringite.

The transition zone is considered the weakest link in the composite chain because of disproportionate porosity and **microcracks**. However, the effect is much more significant because these same voids and microcracks limit transfer of stress to the aggregate, thus weakening the overall composite from the ideal strength. Porosity and microcracks also make the transition zone more permeable to water during the lifetime of the concrete. In steel-reinforced concrete, this increases the susceptibility of the steel to corrosion.

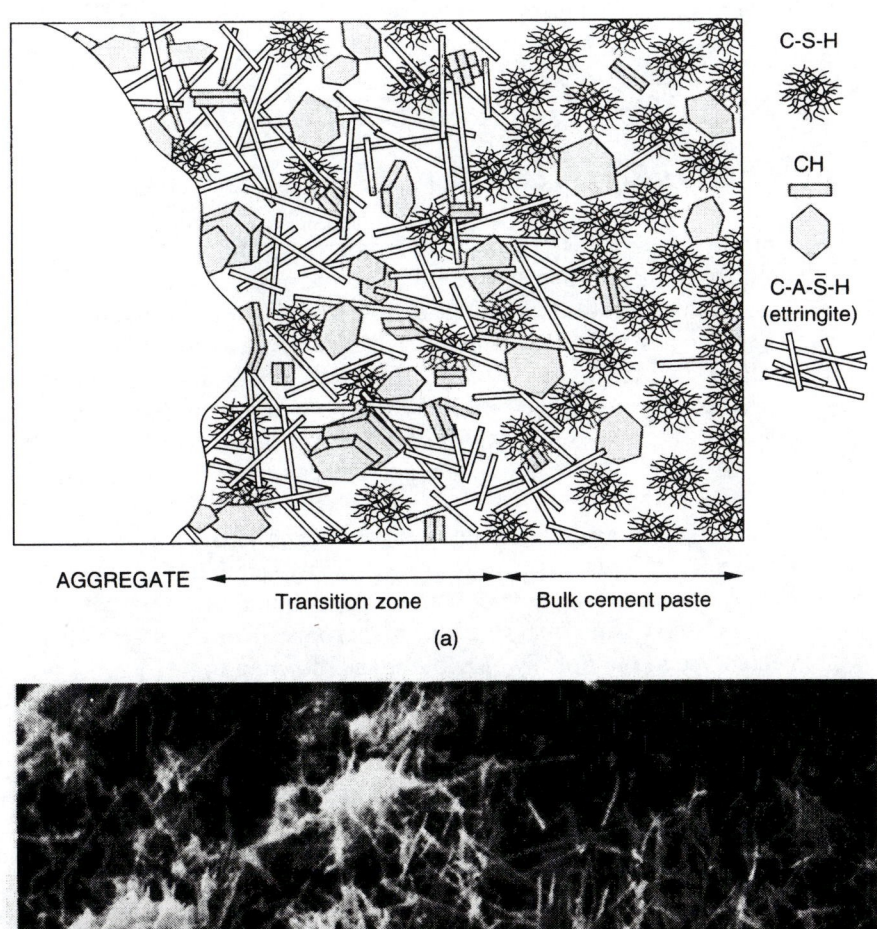

C-S-H

CH

C-A-S̄-H
(ettringite)

AGGREGATE ← Transition zone → ← Bulk cement paste →

(a)

(b)

**Figure 9.4**
Concrete transition zone: (a) schematic diagram, (b) SEM micrograph of
ettringite needles (1500×)
(Part (a) from P. K. Mehta and P. J. M. Montiero, *Concrete: Structure, Properties,
and Materials,* 2nd ed., Prentice Hall, 1993.)

# 9.5   Freeze-Thaw Resistance and Air-Entrained Concrete

Entrapped air voids responsible for porosity and **permeability** to water affect not only strength but also **freeze-thaw resistance**. As water freezes in these pores, it expands and creates hydraulic pressure that can exceed the tensile stress of the concrete. The cumulative effect can cause deterioration in the form of cracking, spalling, and crumbling. One of the greatest advances in concrete technology was the discovery and development of **air-entrainment** to improve the resistance to freezing when exposed to water and deicing chemicals.

Air-entrained concrete is made either with a special cement paste such as Type IA, Type IIA, or Type IIIA or by air-entraining admixtures added during the mixing process. These air-entraining agents enhance and stabilize bubbles formed during mixing, but the resulting voids are extremely small, in the range 10–1000 microns, in comparison to entrapped voids that are 1000 microns or larger and capillary voids in the transition zone, measured in angstroms. Entrained air voids are not interconnected and are uniformly distributed, as shown in Figure 9.5.

Entrained air voids act as empty chambers where excess water expelled by freezing in the concrete can migrate. The magnitude of the hydraulic pressure

**Figure 9.5**
Polished section of air-entrained concrete
(From S. H. Kosmatka and W. C. Panarese, *Design and Control of Concrete Mixtures,* 13th ed., Portland Cement Association, 1988.)

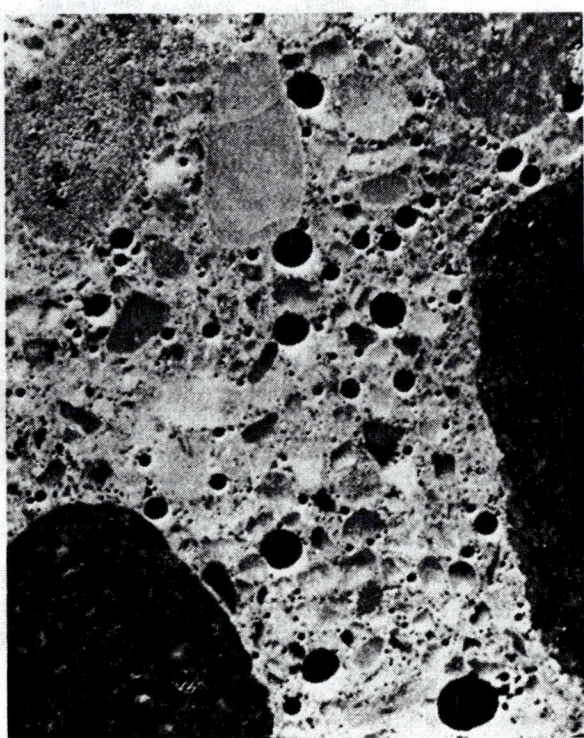

depends on how far the water must travel, the permeability of the material through which it migrates, and the rate at which ice forms.

The recommended entrained air contents depend on exposure, aggregate size, and presence of aggressive chemicals such as sulfates in the water. For moderate exposure and $3/4$-in. aggregate, 5% entrained air is typical. Smaller aggregate size and more severe exposure conditions increase the amount of entrained air. The spacing and size of the entrained voids are also important characteristics that contribute to the effectiveness of entrained air concrete. We should strive for maximum space between nearest voids of 0.008 in. or less. The surface area of voids should be 600 in.$^2$/in.$^3$ of air void volume and the number of voids per linear inch of traverse must be 1.5 to 2 times the percentage of air in the concrete.

# *9.6   Admixtures in Concrete*

An **admixture** is any material that is added to concrete during mixing other than water, aggregates, hydraulic cement, and reinforcement materials. The major purposes of admixtures are for air entrainment, water reduction, reduced segregation, and acceleration or retardation of hydration, among others. We have examined the desirability of air-entraining admixtures in Section 9.5. Actual materials are surface-active, long-chain organic molecules or surfactants with one end of the molecule being *hydrophilic* (water attracting) and the other end being *hydrophobic* (water repelling). The surfactants become adsorbed at the air-water or cement-water interface, where their orientation determines whether they serve to entrain air or increase plasticity.

**Water-reducing** admixtures can be used for several purposes, but not at the same time. For example, tests have proved that a water-reducing admixture can

- double the slump while maintaining the same amount of cement, the same water/cement ratio, and the same compressive strength

- decrease the water/cement ratio and increase the compressive strength for the same slump and cement content

- reduce the cement content by 9% while maintaining the water/cement ratio, slump, and compressive strength

**Superplasticizers** differ from these water-reducing admixtures because they have high molecular weights and are added in much higher amounts than ordinary plasticizers, as much as 1% of the cement weight. Water reductions of the order of 20–25% are frequently possible without reducing the consistency of the fresh concrete. This characteristic is important in building construction and in precast concrete structures (such as Jersey barriers) because high early strengths are needed to reuse the forms as soon as possible.

Admixtures are often used for accelerating or slowing down the hydration of concrete—accelerating to reduce the time for curing in cold weather and slowing to extend the time for curing in hot weather. Retarding admixtures also keep the concrete workable throughout the entire time needed for placing and finishing. The most common accelerating admixture is calcium chloride. When added in amounts from 0.5% to 2% of the cement weight, this admixture reduces time for curing, increases early strength, and increases heat of hydration. Its use is not recommended for reinforced concrete, however, because of the corrosion hazard posed by the chloride content.

When using retarding admixtures, the temperature of the mix water is always lowered. The admixtures, which can be carbohydrates or organic acids, do not lower the temperature of the mix. Instead, they offset the accelerating effect of hot weather on the initial set, thereby assisting placement and finishing operations. Most retarding admixtures are also water reducing, thereby providing higher strengths as well as the improved workability.

# 9.6.1 Mineral Admixtures

Mineral admixtures are finely divided powders that are added in relatively large amounts, in the range 20–90% of the cement content of concrete mixes. These are classified as cementitious materials, pozzolans, or materials with both cementitious and pozzolanic characteristics.

A **pozzolan** is a siliceous or aluminosiliceous glassy (noncrystalline) material that in itself is not cementitious, but in finely divided form reacts with water and calcium hydroxide to form compounds that have cementitious properties. Pozzolans from the volcanic ash of Vesuvius were used as **mortar** in the aqueducts built by the ancient Romans. Many pozzolans occur in nature and must be crushed, ground, and thermally treated before they can be used. The most widely used mineral admixtures in concrete, however, are industrial process by-products, such as low calcium fly ash and silica-fume particles, shown in Figures 9.6 and 9.7. Low-calcium **fly ash** is the by-product of electric power generation through combustion of pulverized coal. The mineral impurities in the coal do not burn and are carried away with the exhaust gases, where they are collected as fine glassy spherical powders by electrostatic collectors in the stacks. Condensed silica fume is a by-product of the silicon and ferrosilicon industries where $SiO$ vapors are formed during the reduction of quartz to silicon; these vapors oxidize and condense at low temperatures to extremely fine particles that have high surface area-to-volume ratios.

**Cementitious** materials are those that are true hydraulic cements. They include ground **blast furnace slag**, produced as a by-product in steelmaking, and hydrated lime, made by calcining limestone containing silica and alumina. Other materials, such as high-calcium fly ash produced by burning lignite or subbituminous coals, have a combination of cementitious and pozzolanic properties.

Concretes that contain these fine mineral admixtures will generally require less water for equivalent workability than mixes having only portland cement. This is the major reason for the use of these admixtures, but there are other benefits as well. They reduce the segregation and heat of hydration and retard the setting time.

**Figure 9.6**
SEM micrograph of fly ash
particles (1000×)
(From S. H. Kosmatka and
W. C. Panarese, *Design and
Control of Concrete Mixtures,*
13th ed., Portland Cement
Association, 1988.)

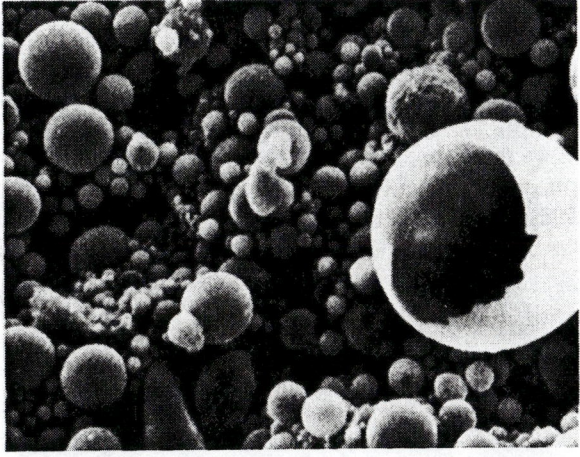

**Figure 9.7**
SEM micrograph of silica
fume particles (20,000×)
(From S. H. Kosmatka and
W. C. Panarese, *Design and
Control of Concrete Mixtures,*
13th ed., Portland Cement
Association, 1988.)

Higher strengths are the usual case, but the development of the strength takes longer. They generally reduce the permeability of the cured concrete, thereby reducing corrosion problems for steel reinforcement.

# 9.7 Curing of Concrete

Curing is synonymous with hydration, a time-dependent process that continues as long as water is available and unreacted cement remains. As curing progresses, concrete becomes stronger, as shown in Figure 9.8, more **abrasion resistant**, more

**Figure 9.8**
Effect of curing conditions and
time on compressive strength
of concrete
(From S. H. Kosmatka and
W. C. Panarese, *Design and
Control of Concrete Mixtures,*
13th ed., Portland Cement
Association, 1988.)

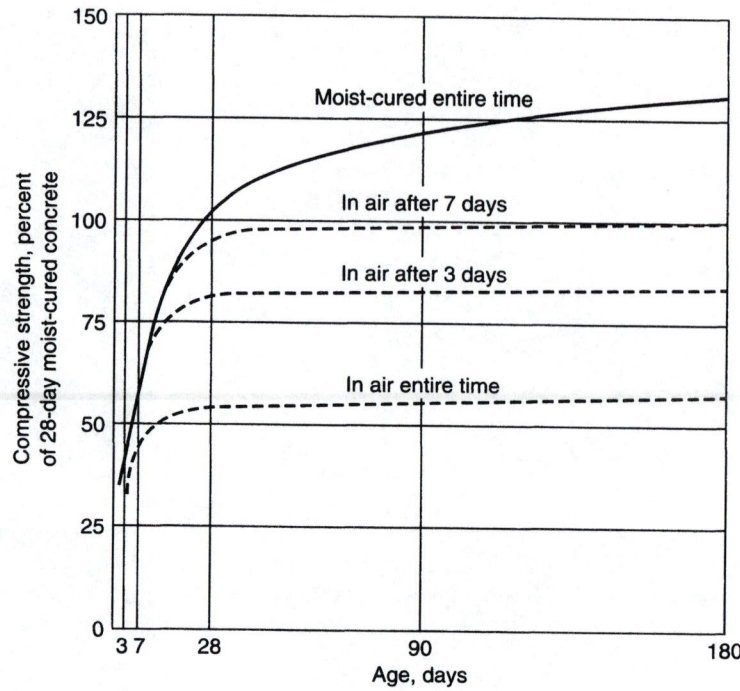

impermeable, and more resistant to freeze-thaw cycles. There is considerably more than enough water in fresh concrete to complete hydration, but any appreciable evaporation or other water loss can delay or prevent complete curing. In addition, the hydration is temperature dependent, so any unfavorably cold conditions ($T<50°F$) at the early stages of curing can be detrimental. Therefore, we must prevent or replace any loss of moisture from the concrete and maintain favorable temperatures during the curing process.

The three methods that we can use for maintaining water during curing are

1. maintenance programs that include spraying, saturated covers, and even immersion

2. loss prevention whereby concrete is covered with plastic sheets, impervious papers, or liquid membrane–forming compounds in order to minimize evaporation

3. acceleration through heating and providing additional moisture, usually by steam curing

The curing period during which concrete should be protected from evaporation depends on the type of cement, the concrete mix, mass of the structure, weather conditions, and required strength. For large structures, such as dams, this period might be as long as 3 weeks, but the rule-of-thumb is the time to develop 70% of the specified compressive strength. This is typically 7 days for Type I cement concrete, 3 days for Type III cement concrete, and 24 hours for steam curing.

### Case Study 9.2

#### A Cold-Weather Concreting Problem

RDX Corporation, near the University of Massachusetts Lowell, is a sprawling research center with eight multistory buildings. In order to comply with recent regulations for handicapped access, RDX contracted to have access ramps constructed for all sidewalks in the complex. It was December before the contractor was able to complete the forms and pour the concrete. The fresh concrete was covered with a plastic sheet, then with straw, but this proved to be insufficient. By springtime, the edges were spalling, with powder and small pieces of concrete collecting below the curb. Concerned with safety, RDX management asked the Civil Engineering Department at the university to determine the cause of the spalling.

Spalled sections of the concrete were provided for examination in a scanning electron microscope. The fracture surfaces, shown in Figure 9.9, are covered with a calcium crystalline deposit, unlike fracture surfaces of any properly cured concrete. It was concluded that the heat loss at the edge of the freshly poured concrete was sufficient to cause freezing that interfered with curing. As a result of this simple study, the contractor reluctantly agreed to patch the curb sections.

# 9.8    Reinforced Concrete

The strength of concrete in tension is very low in comparison to its compressive strength. This is a result of the stress concentration of the numerous microcracks and entrapped air formed during curing. Steel has good tensile strength but is expensive by comparison and corrodes as well. **Reinforced concrete** combines the compressive strength of concrete and the tensile strength of steel for many structural applications. Not only is this an economical approach, but the two materials are well suited to each other, having matching thermal expansion coefficients and good bonding characteristics due to the shrinkage of the concrete onto steel reinforcement during curing.

## 9.8.1    Design Considerations

The shapes for which reinforcement is used include beams, columns, and slabs, and the reinforcement must be placed where tensile stresses are encountered during use. Designers must also be concerned with shear stresses in use, but design is not the focus of our studies. We are concerned, however, with the strength of the reinforced concrete members, which depends mostly on the transfer of tensile stress from the concrete to the steel.

**Figure 9.9**
SEM micrographs of fracture surface of spalled edge of concrete curb at RDX Corporation: (a) overall surface condition (100×), (b) crystalline surface formation (1000×)

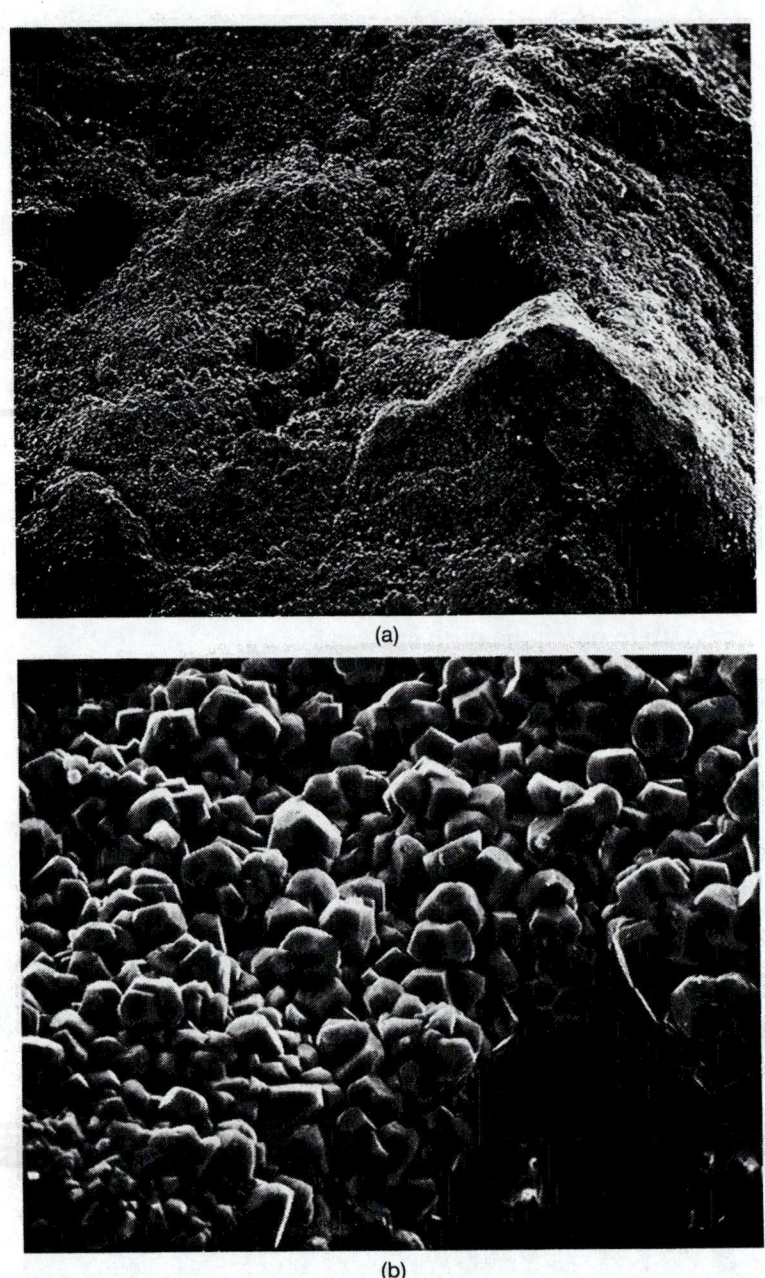

(a)

(b)

Although the primary bond strength is derived from the concrete shrinkage onto the steel during curing, there are designs that we can use to improve these bonds between concrete and steel even more. For example, bending the ends of the steel reinforcement helps to prevent any slipping along the length where tension is greatest. We can also use reinforcing bars, called **rebars**, that have raised ribs to promote bonding, or we can use prefabricated welded fabric mesh where crossing wires prevent any slipping of the wires in tension.

## 9.8.2    *Prestressed Concrete*

Materials that have poor tensile strength can be improved by building compressive stresses into them before they are placed into service. For example, we have seen the advantages of this in tempered glass. The method is also used successfully for pre-cast concrete beams. Continuous steel prestressing strands are placed in tension while concrete is poured around them and allowed to cure. When the tensile load is removed, the strands contract, placing the concrete now bonded to them in compression. When in use, a tensile force equivalent to this compressive force can be applied and the **prestressed concrete** will only return to the unstressed condition; it will not see any tensile force. Once at this stage, it will behave as "ordinary" unreinforced concrete under additional tensile forces.

Post-tensioning is also practiced. In these cases, however, the tensioning strands are lubricated so that they will not be bonded to the concrete. Once curing takes place, the strands are loaded in tension and the concrete and strands are placed between blocks (affixed to the strands). When tension is released, the blocks place the concrete in compression. This technique is used in many highway bridges.

When prestressed concrete is used, it makes economic sense to combine prestressed units that are made more readily in the factory with the concrete poured in place. This is advantageous because of the difficulties of prestressing on-site and because only a small part of the structure actually has to be prestressed.

## 9.8.3    *Corrosion of Reinforcing Steel*

Corrosion is a major factor in failure of reinforced concrete. Although concrete is itself water-resistant, it is permeable to water and corrosion cells can easily be established. This is particularly true where rebars do not have sufficient cover and freeze-thaw conditions can easily lead to spalling of the concrete, thus exposing the rebars, and where chlorides are present such as in marine environments (recall Case Study 9.1). Rebars are available with epoxy coatings intended to protect the rebars from corrosion. However, bending the rebars to prevent slipping can cause breaks in the epoxy, enhancing corrosion because of the large cathode-to-anode area ratio. Other materials, such as Kevlar, are being considered for replacement of steel as well.

## 9.8.4    *Fiber Reinforcement*

We learned about **fiber reinforcement** in studying composite materials. Early investigations into the potential of fiber reinforcement of concrete did not provide strengthening, mainly because workability was reduced so much that the volume fraction of fibers that could be introduced was insufficient to contribute to overall strength. However, the studies did show dramatic improvement in the behavior after cracking occurred. Although strength was not affected, the tensile strain to rupture was markedly improved. Thus, fiber-reinforced concrete displays improved toughness when compared to ordinary concrete. This is attributed to the bond strength because the pullout strength exceeds the load when cracking first occurs. Fibers transfer the stress to uncracked concrete until either the fibers fail or fiber pullout occurs because of cumulative cracking.

Steel fiber concentrations of 1–2% by volume have produced 15–30% increase in the flexural strength of fiber-reinforced mortars. Improvement in toughness has been related to fiber concentration and resistance to pullout, which in turn is related to the length-to-diameter, or aspect, ratio of the fibers.

Although many fiber materials are available, steel fibers predominate because they are available with aspect ratios of 30 to 150, are inexpensive, and are ductile, so ends can be crimped to enhance pullout strength. Typical fiber-reinforced concrete is used for paving highways and airport runways and for overlays where it is specified by flexural strength (700–900 psi) and by fiber content (0.38–2.0% by volume). Because the workability is reduced by the fibers, we usually increase the ratio of fines to coarse aggregates and incorporate fly ash into the mix.

## 9.9    *Polymer Concrete*

Impermeable concretes with very high compressive strengths can be produced using polymers instead of or in addition to portland cement. Although the technology has been available for many years, the high cost has limited applications to emergency repairs where short setting times, within a few hours at most, are imperative. **Polymer concrete** (PC) is made using only a monomer as the bonding agent, whereas latex-modified concrete (LMC) incorporates a latex polymer in place of part of the water in a normal concrete mix. Polymer impregnated concrete (PIC) uses precast portland cement concrete impregnated with a liquid or gaseous monomer that polymerizes in situ. The improvement in mechanical properties for these concretes incorporating polymers is demonstrated in Table 9.4.

**Table 9.4**
Mechanical properties of polymer concretes

| Concrete type | Polymer | Compressive strength (psi) | Tensile strength (psi) | Flexural strength (psi) |
|---|---|---|---|---|
| PC | Polyester (1:9 P:Agg) | 18,000 | 2000 | 5000 |
| | MMA (1:15 P:Agg) | 20,000 | 1500 | 3000 |
| LM mortar | Control | 4500 | 39 | 69 |
| | Styrene/ Butadiene* | 4800 | 830 | 1430 |
| | Acrylic* | 5700 | 835 | 1835 |
| PIC | Control | 4950 | 335 | 630 |
| | MMA** | 20,250 | 1630 | 2640 |
| | Styrene** | 14,140 | 190 | 2300 |

\* Polymer-cement ratio = 0.2
\*\* Approximately 5 weight percent polymer loading

## Case Study 9.3

### More Corrosion Problems

Sometimes we take a new product to market too soon, trying to be first with the best. Perhaps we should reconsider after pondering what went wrong with a 23-story bank building in Cleveland some years ago. Architects were fascinated by the high strength of mortar made with vinylidene chloride latex monomer. Compressive strengths of more than 8000 psi, tensile strengths of almost 1000 psi, and flexural strengths over 1800 psi were typical of this mortar, making possible a freestanding single-course brick wall, known as a Miller wall. The cost saving in brickwork alone was remarkable!

Attitudes change, however, when our ideas go sour. Nine years after construction, the walls began cracking so badly that immediate repair was necessary and the total bill was more than $5 million! The cause? It didn't take long to determine that cracking and displacement was the result of excessive rust scale on the steel members of the building. The rust contained significant amounts of chloride that leached from the vinylidene chloride. Don't blame Cleveland, though. The problem was repeated many times across the country, leading the manufacturer to discontinue sales of the latex modifier.

# 9.10   Design Control of Cracking _____

Formation of large cracks in concrete can be unsightly at best. Cracking can be caused by stresses arising from applied loads, from drying shrinkage, or from thermal changes. Drying shrinkage is unavoidable, but cracking caused by the shrinkage can be reduced by properly positioned rebars or by designing control joints that direct cracks to inconspicuous locations. These **design joints** are grooved, scored, or sawed in sidewalks, driveways, pavement, and floors. Two other types of design joints are also used—isolation joints that separate a slab from other parts of a structure, allowing both horizontal and vertical movement of the slab, and construction joints, which separate areas of concrete placed at different times. Figure 9.10 illustrates these joint designs.

# Summary _____

Concrete, both a ceramic and a composite, is one of the world's most common construction materials. It is made of a mixture of coarse and fine aggregates, water, and portland cement, a mixture that hydrates over a period of time to form a solid that is strong in compression but weak in tension because of entrapped air voids and microcracks. Hydration is a complex chemical reaction of the components in cement with water and depends to a large extent on the composition of the cement. The major contribution to strength and bonding to the aggregates comes from calcium silicate hydrates, C-S-H, which are poorly defined crystals with very high surface area/volume ratio. Concrete mixtures are based on workability, aggregate size, and water/cement ratio; the latter directly affects the strength that can develop with hydration. Admixtures are added to entrain air for improved freeze-thaw resistance, to improve workability, and for reduced water content of concrete mixes. Large amounts of finely ground mineral admixtures, which are pozzolanic or cementitious, also reduce water requirements for workability. Curing is most important in development of strength in concrete. Practical methods control the water needed for hydration by adding water, covering to prevent loss, or by accelerating hydration. Many applications utilize reinforced steel to enhance the tensile properties of beams, columns, and slabs. Reinforcement can also prestress the concrete to promote further enhancement of apparent tensile strength and fiber reinforcement can increase the tensile strain before fracture. Finally, polymers can also strengthen concrete substantially, but high cost limits their use.

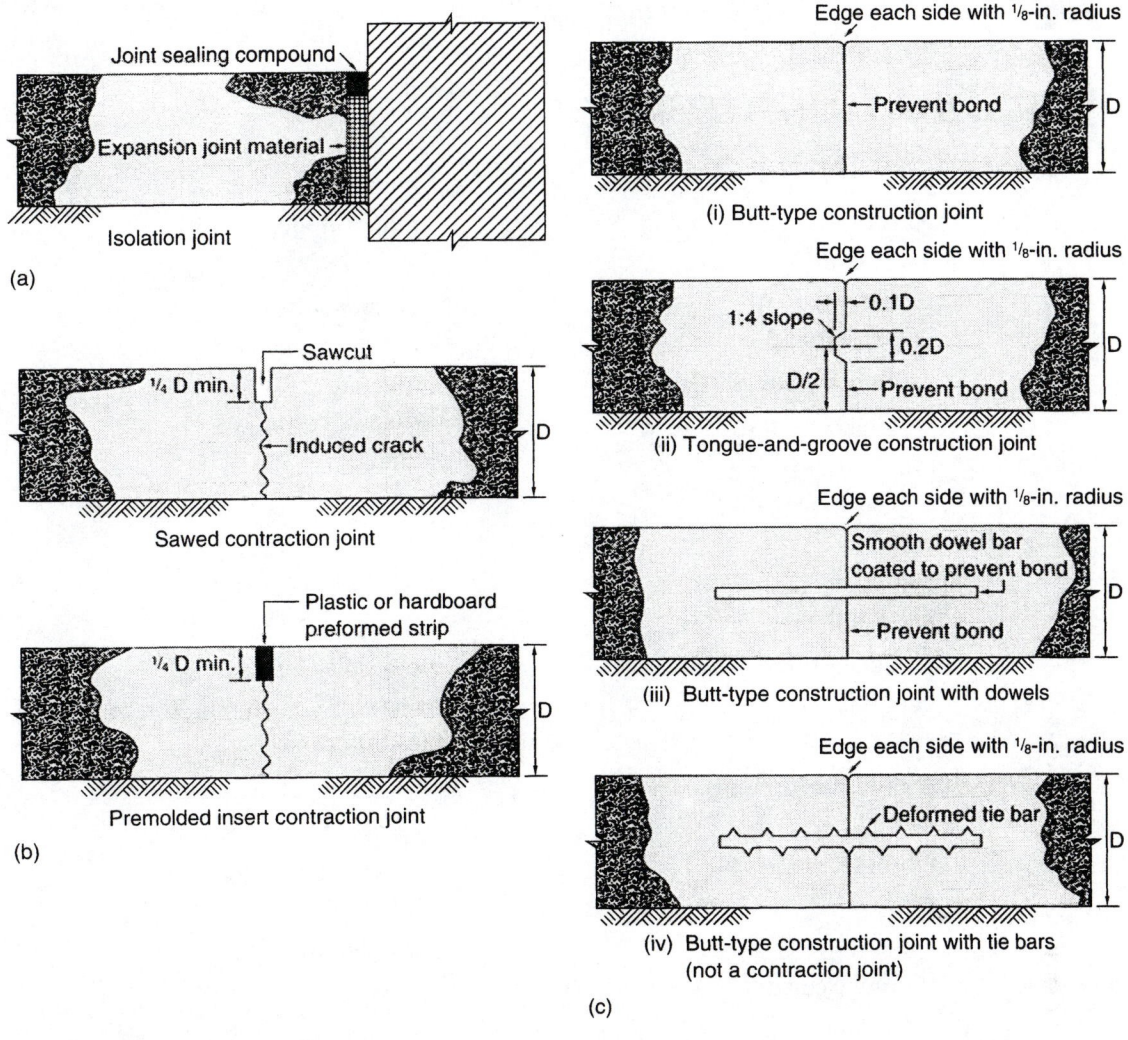

**Figure 9.10**
Concrete design joints: (a) Isolation joints permit horizontal and vertical movements between abutting faces of the slab and fixed parts of a structure. (b) Contraction joints provide for horizontal movement in the plane of a slab or wall and induce controlled cracking caused by drying shrinkage. (c) Construction joints are stopping places in the process of construction. Construction joint types i, ii, and iii are also used as contraction joints.
(From S. H. Kosmatka and W. C. Panarese, *Design and Control of Concrete Mixtures,* 13th ed., Portland Cement Association, 1988.)

# Terms to Remember

abrasion resistance

admixture

aggregate

air-entrainment

blast furnace slag

calcium hydroxide (CH)

calcium silicate hydrate (C-S-H)

cementitious

compressive strength

concrete

curing

design joints

dicalcium silicate (C₂S)

durability

ettringite

fiber reinforcement

fly ash

freeze-thaw resistance

gypsum

hydration

hydraulic cement

microcracks

monosulfate

mortar

permeability

polymer concrete

porosity

portland cement

pozzolan

prestressed concrete

rebars

reinforced concrete

slump

superplasticizers

tetracalcium aluminoferrite (C₄AF)

transition zone

tricalcium aluminate (C₃A)

tricalcium silicate (C₃S)

water/cement ratio

water-reducing

workability

# Problems

1. Describe in your own words the role of aggregates in determining the properties of concrete, including what materials are used as aggregates, what factors affect their selection, and what preparation is necessary.
2. Describe in your own words the process of curing, or hydration, of portland cement.
3. Using Sample Problem 9.1 as a guide, determine the batch weights for concrete with the following properties:
   a. slump is 2 in.
   b. maximum aggregate size is 1 in.
   c. 28-day compressive strength is 5000 psi

4. Explain why C-S-H is more important than ettringite in determining the final strength characteristics of concrete.
5. What is freeze-thaw resistance and how can it be altered?
6. What are admixtures? Explain the benefits of two different types of admixtures and how they might influence the final properties of concrete.
7. Describe in your own words practical ways that water is controlled during curing of concrete.
8. How can we prevent or control cracking of concrete during curing? Is the effect only cosmetic?
9. Explain in your own words the benefits of concrete reinforcement.
10. Explain in your own words the benefits of fiber reinforcement of concrete.

# 10

# *Wood and Wood Products*

Although we are all familiar with **wood** and wood products, we don't ordinarily think of them as industrial materials. Yet they are used in every industry from construction, where more wood is consumed every year than steel, to publishing, which uses many grades of paper made from wood. Products from packaging to material handling to structural timbers made from wood are used and reused in almost all industries. In fact, wood is one of the earliest materials used by humans, who used it to build fires and make some of the world's first tools.

Wood is very different from all other materials we have studied because it was at one time alive. We separate wood into two types—**hardwood** that comes from **deciduous** trees, such as oak or maple, and **softwood** that comes from **conifers**, that is, cone-bearing trees. Both types convert carbon dioxide and moisture into **cellulose**, **lignin**, and hemicellulose by the process of photosynthesis, which takes place in the **cambium** layer separating the bark and outermost **sapwood**. In both hardwood and softwood, older sapwood becomes converted to **heartwood** and the diameter of the trunk increases as new sapwood is formed. Because of this growth pattern, the properties of wood lumber depend on its orientation within the trunk it came from. In addition, all trees branch outward to crown the trunk, thus increasing the surface area for photosynthesis, but the branching process produces knots in lumber, which introduces even more inhomogeneous character to the wood.

We will examine the structure of wood in this chapter and its effect on properties, but we will also find that other factors, such as **moisture content**, also have an influence on these properties. We will look at methods to preserve and protect wood from molds, decay, and insect infestation. Finally, we will examine some of the many wood products such as plywood, particleboard, and paper that have found many applications in construction, packaging, and documentation.

# 10.1   Structure of Wood

The structure of wood is complicated not only by the growth process but by the many species of both softwoods and hardwoods. It is common knowledge that the age of a tree can be established accurately by the **growth rings** that are caused by the seasonal growth patterns. In spring when the sap is running and new buds are forming, growth is rapid. The cells produced are less dense (large diameter and thin walls) because they must carry the food and water from the roots to the leaves and back. Later, in the dry summer season, growth is slowed and denser cells (small diameter and thick walls) are formed because there is less food transported. The resulting cross section that is produced is shown in Figure 10.1. In this figure, *pith* is the soft material at the center of the tree where the first growth took place.

When lumber is cut from logs, the length of the boards is parallel to the fibrous cell growth, but the pattern (and, therefore, structure) of the width and thickness is dependent on where the board was cut in the cross section of the log. The orientation is described by the convention given in Figure 10.2. The characteristics of each cut will affect shrinkage in drying, strength, and the grain or appearance.

Wood cells that form the wood microstructure differ for softwoods and hardwoods. In softwoods, cells are mainly fibrous, with individual **fibers**, called **tracheids**, bonded to each other by lignin. Tracheids are typically 3–8 mm in length and serve to transport the sap in softwoods. In hardwoods, however, cells are much more varied in size and have a small length-to-diameter (aspect) ratio. The softwood tracheids and segments that make up **vessels** in hardwoods are shown in Figure 10.3. Included in this figure is a **libriform** fiber that is also found in hardwoods. These fibers, although more prevalent than vessel segments, are more important structurally than for conduction of sap. Sap movement in hardwoods occurs mainly in the vessels that are surrounded by the libriform fibers and vessel segments and that are many times larger than the cells. Although most cells are oriented longitudinally, both hardwoods and softwoods have some cells that are oriented radially. These are called **rays** and serve to conduct sap radially across the grain.

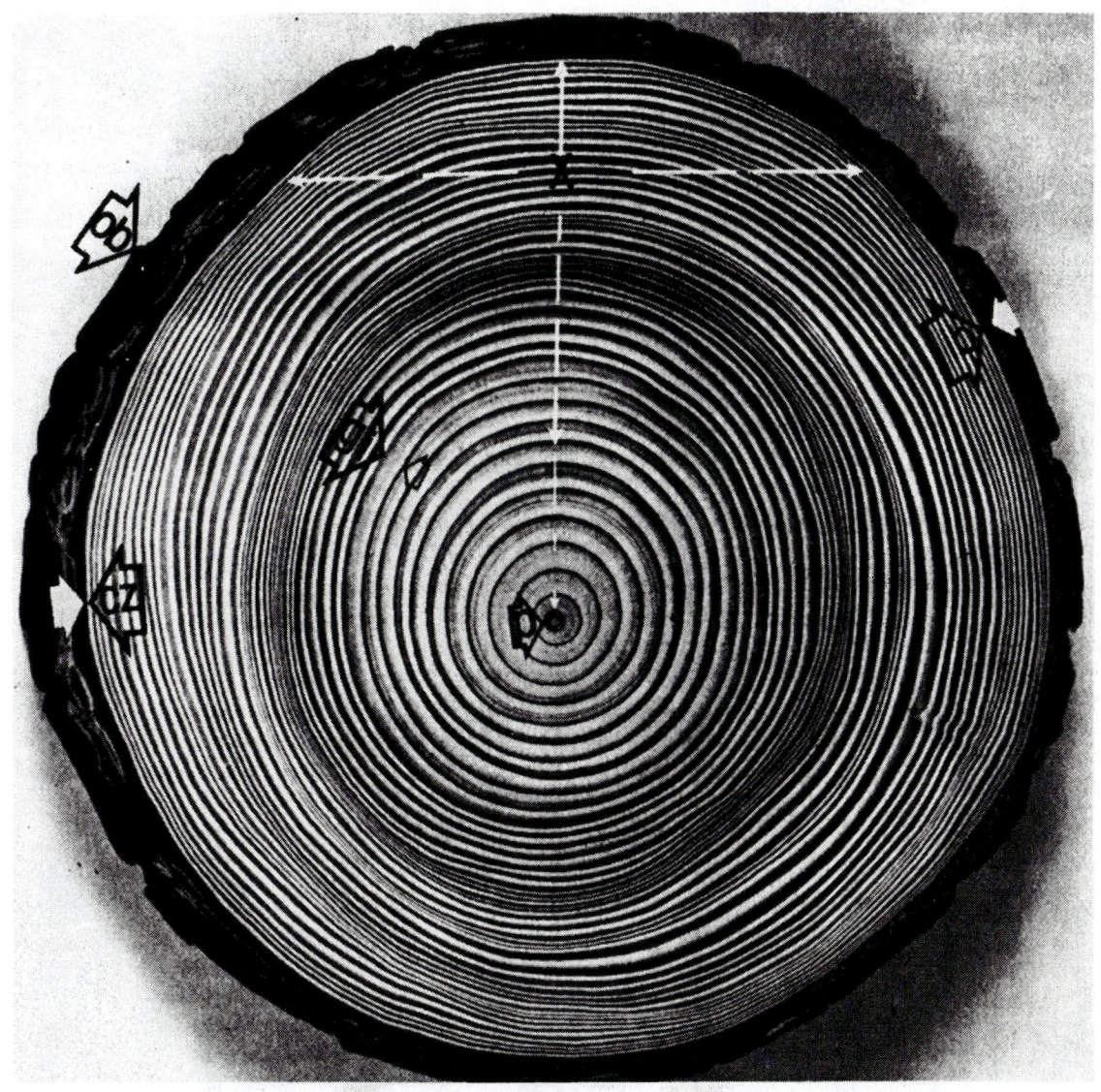

**Figure 10.1**
Cross section of the trunk of a Douglas fir. Key: cz = cambial zone, p = pith, gi = growth
ring, x = xylem (wood), ob = outer (dead) bark, and ib = inner (live) bark.
(From H. A. Core, W. A. Côté, and A. C. Day, *Wood Structure and Identification,* 2d ed.
(Syracuse: Syracuse University Press, 1979). By permission of the publisher.)

**Figure 10.2**
The principal directions of wood with respect to grain and growth ring orientation

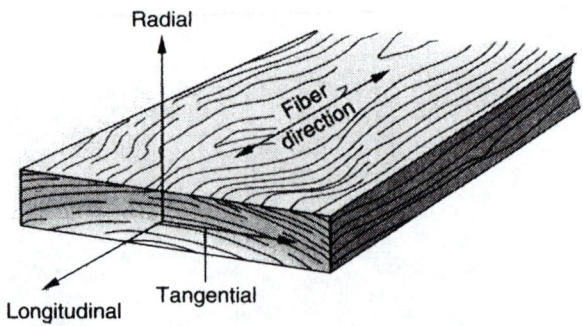

Radial

Fiber direction

Longitudinal

Tangential

**Figure 10.3**
Comparative sizes of cell types in hardwoods and softwoods: *a*, *b*, and *c* are vessel segments; *d* is a libriform fiber; and *e* is a tracheid (the softwood tracheid is about 3.5 mm long).
(From H. A. Core, W. A. Côté, and A. C. Day, *Wood Structure and Identification,* 2d ed. (Syracuse: Syracuse University Press, 1979). By permission of the publisher.)

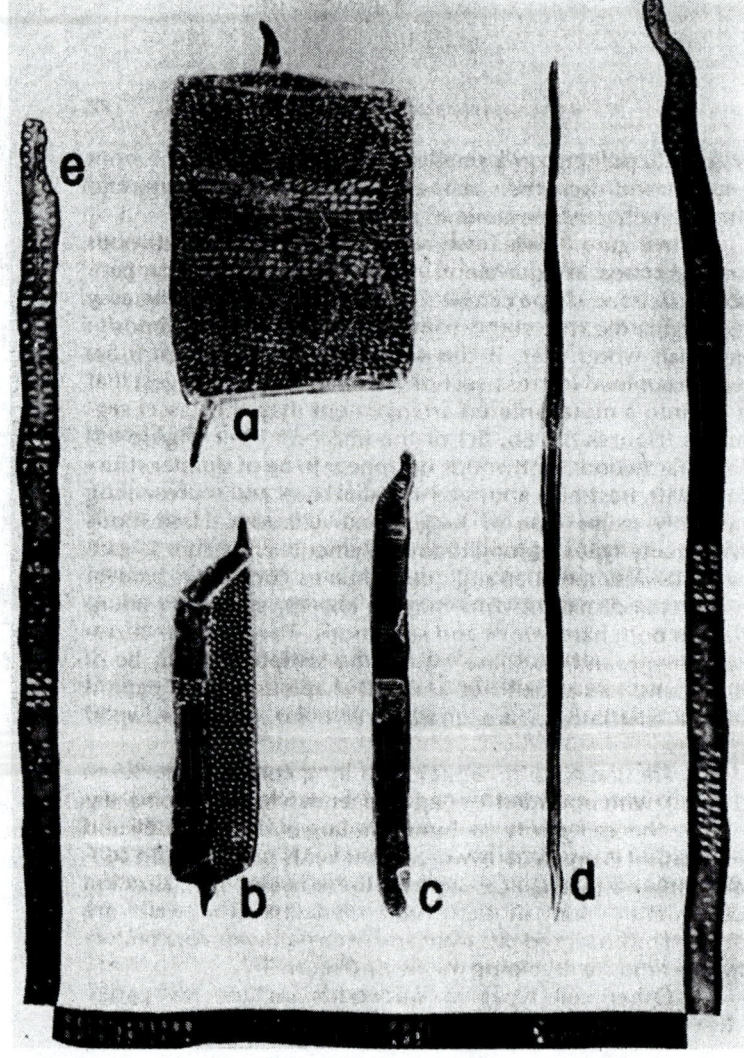

In softwoods, the tracheids formed early in the growing season are large and thin-walled. As the season changes, however, the cell walls become thicker and the cell diameter diminishes. The change can be gradual or abrupt, as shown in Figures 10.4 and 10.5. Note the **resin canals** in these figures. Resin is one of the extraneous components that can be present in wood and is valuable when extracted for use in fluxes for metal joining. The presence or absence of resin canals can also be helpful in identifying the species of wood.

The change from **earlywood** to **latewood** in hardwoods can also be abrupt or gradual. Those species in which large differences in the vessel diameter occur are called *ring-porous* to differentiate them from *diffuse-porous* species in which few differences in the vessels occur. Figures 10.6 and 10.7 illustrate these characteristics for red oak (ring-porous) and sugar maple (diffuse-porous).

**Knots** occur in the structure of lumber because of tree limbs growing from the trunk. The cells of the limb are bonded to the cells of the trunk just as they are to other cells within the limb. Knots that result are called *intergrown* knots. For many softwood trees, lower limbs die and there are no connective cells. The trunk eventually grows around the dead area, producing an *encased* knot. The appearance of knots is dependent on how the wood is cut from the log. If the cut is tangential to the growth rings, it is called a plain cut and any knots usually appear to be round. If, however, the growth rings are 45–90° to the board width, then it is called a quarter cut and spike knots frequently are observed. Several types of knots are shown in Figure 10.8. We will see that knots limit the mechanical properties of wood because of grain distortion around the knot and stress-concentration factors.

Whereas trees are living long before they become lumber, there are certain changes in the structure of wood produced by stresses the tree encounters while alive. We call this unusual wood *reaction wood* because it is believed to be caused by the reaction of the tree to excessive stress. The reaction wood occurs from the compression due to weight of a limb or leaning trunk. In softwoods, the change in wood structure occurs at the compression side (bottom) and the reaction wood is called *compression wood*. In hardwoods, the change in structure occurs at the tensile side (top) and the reaction wood is called *tension wood*.

# 10.2  Properties of Wood

The **anisotropy** of the wood structure is manifested in almost all wood properties. The structure alone, however, is not the only factor that controls properties. For example, trees have moisture present in them. We call the natural moisture content of a newly cut tree the *green* condition, which can vary from 30% to 200%, depending on species, sapwood, or heartwood and time of season. However, the structure of wood is **hygroscopic**, so water can also be absorbed unless treated. Moisture content, whether green or absorbed, affects mechanical, electrical, and chemical properties. Another factor that affects properties is the density, or specific gravity, that is

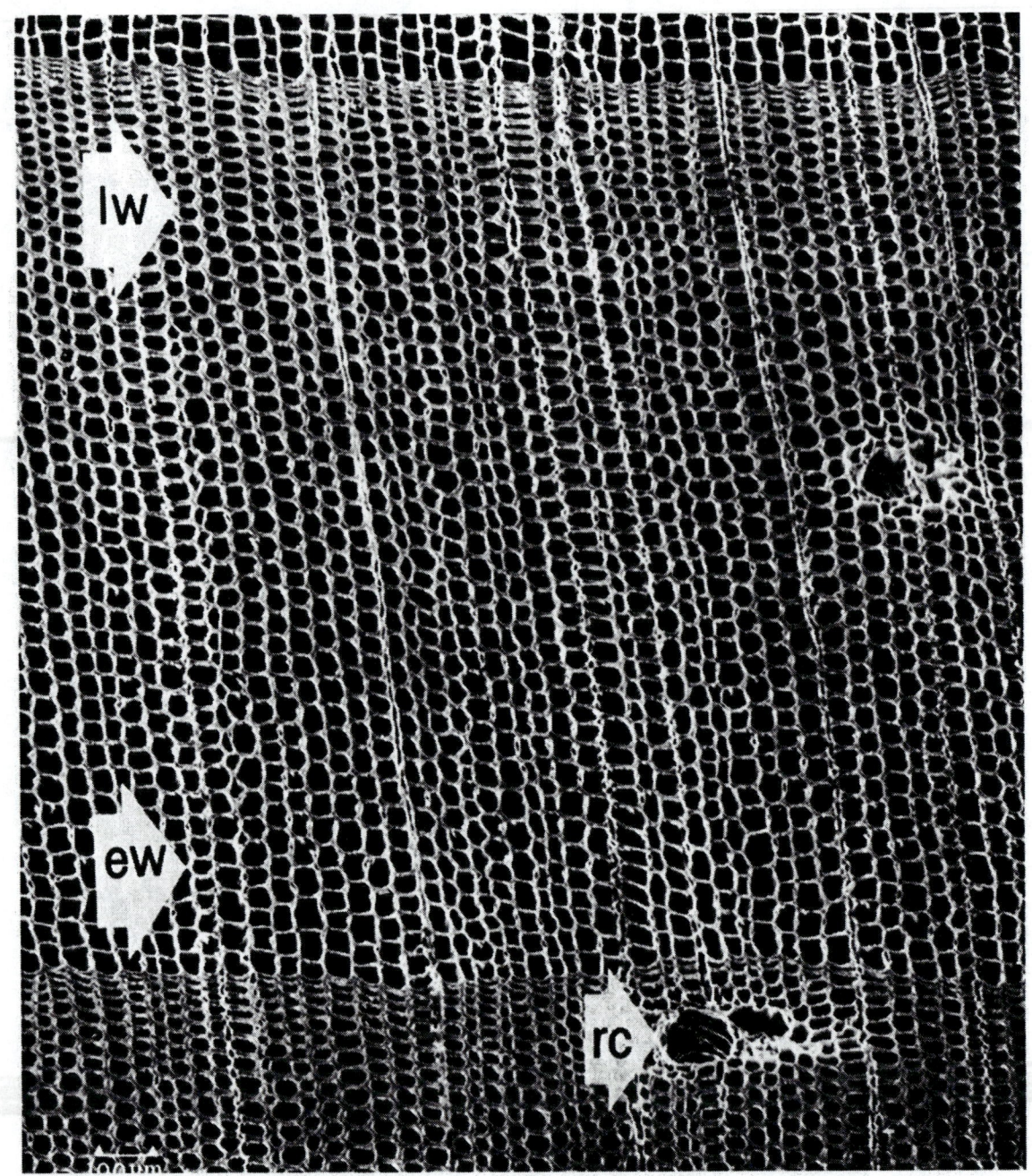

**Figure 10.4**
SEM micrograph showing gradual transition in a coniferous wood, eastern white pine.
Tracheid size changes gradually from earlywood (ew) to latewood (lw). Note the longitu-
dinal resin canals (rc) on this transverse surface.
(From H. A. Core, W. A. Côté, and A. C. Day, *Wood Structure and Identification,* 2d ed.
(Syracuse: Syracuse University Press, 1979). By permission of the publisher.)

**Figure 10.5**
SEM micrograph showing abrupt transition in a coniferous wood, red pine. Tracheid size changes abruptly from earlywood (ew) to latewood (lw). Note the resin canals (rc) on the transverse surface and the ray (r) on the radial surface.
(From H. A. Core, W. A. Côté, and A. C. Day, *Wood Structure and Identification,* 2d ed. (Syracuse: Syracuse University Press, 1979). By permission of the publisher.)

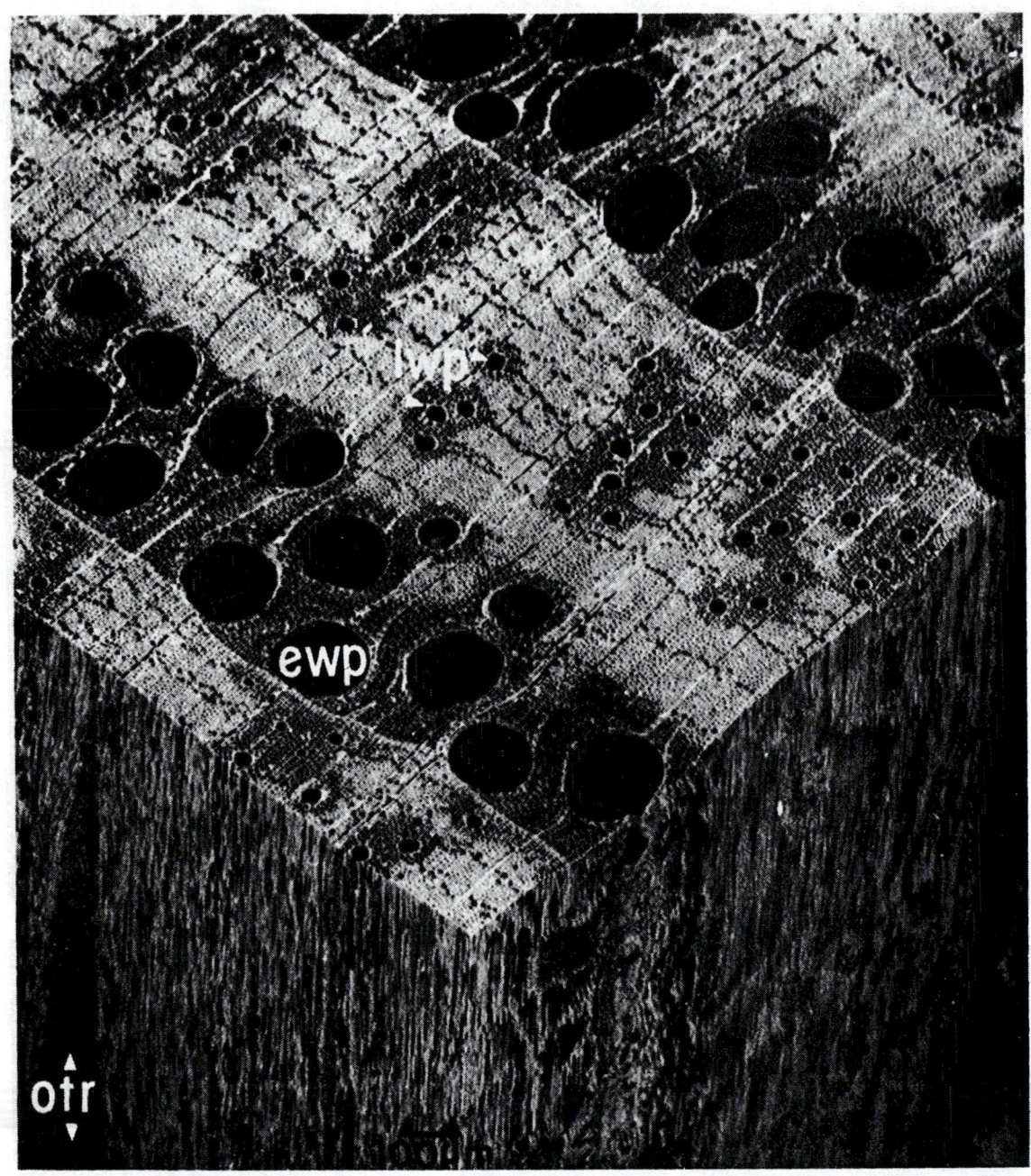

**Figure 10.6**
SEM micrograph of red oak, illustrating a ring-porous growth ring pattern. The spring-wood or earlywood pores (ewp) are large and concentrated in a narrow band at the beginning of the year's growth. Summerwood or latewood pores (lwp) are much small-er. Note the broad oak-type ray (otr) on the tangential surface of the block.
(From H. A. Core, W. A. Côté, and A. C. Day, *Wood Structure and Identification,* 2d ed. (Syracuse: Syracuse University Press, 1979). By permission of the publisher.)

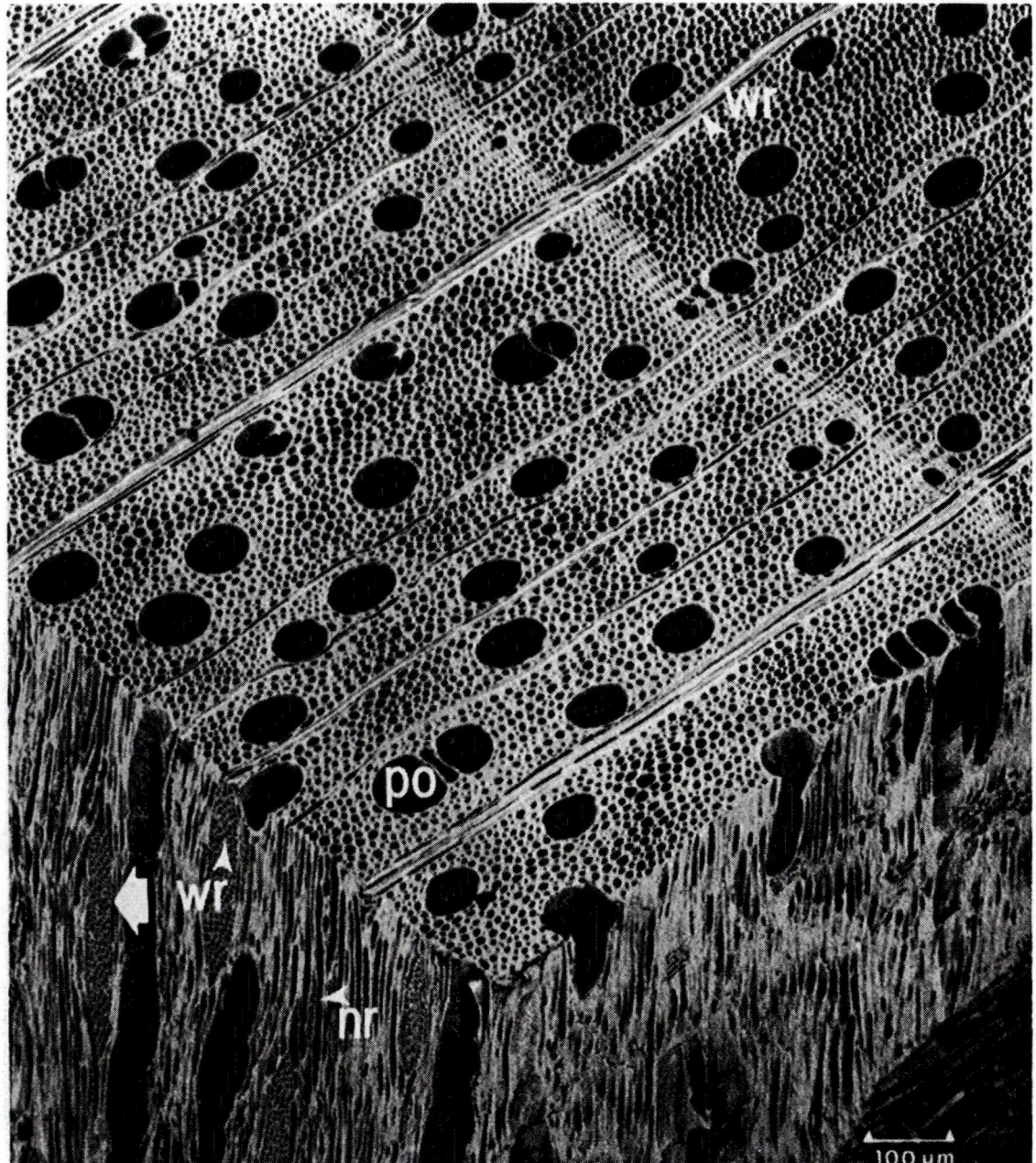

**Figure 10.7**
In a diffuse-porous growth ring pattern, the vessels are more nearly uniform in size
throughout the growth ring than in either ring-porous or semi-ring-porous patterns. The
pores (po), or vessels viewed in cross section, are generally distributed rather uniformly
as well. This SEM micrograph of sugar maple reveals that two sizes of rays are found in
this species, the larger rays being as wide or wider than the pores (wr = wide ray, nr =
narrow ray).
(From H. A. Core, W. A. Côté, and A. C. Day, *Wood Structure and Identification,* 2d ed.
(Syracuse: Syracuse University Press, 1979). By permission of the publisher.)

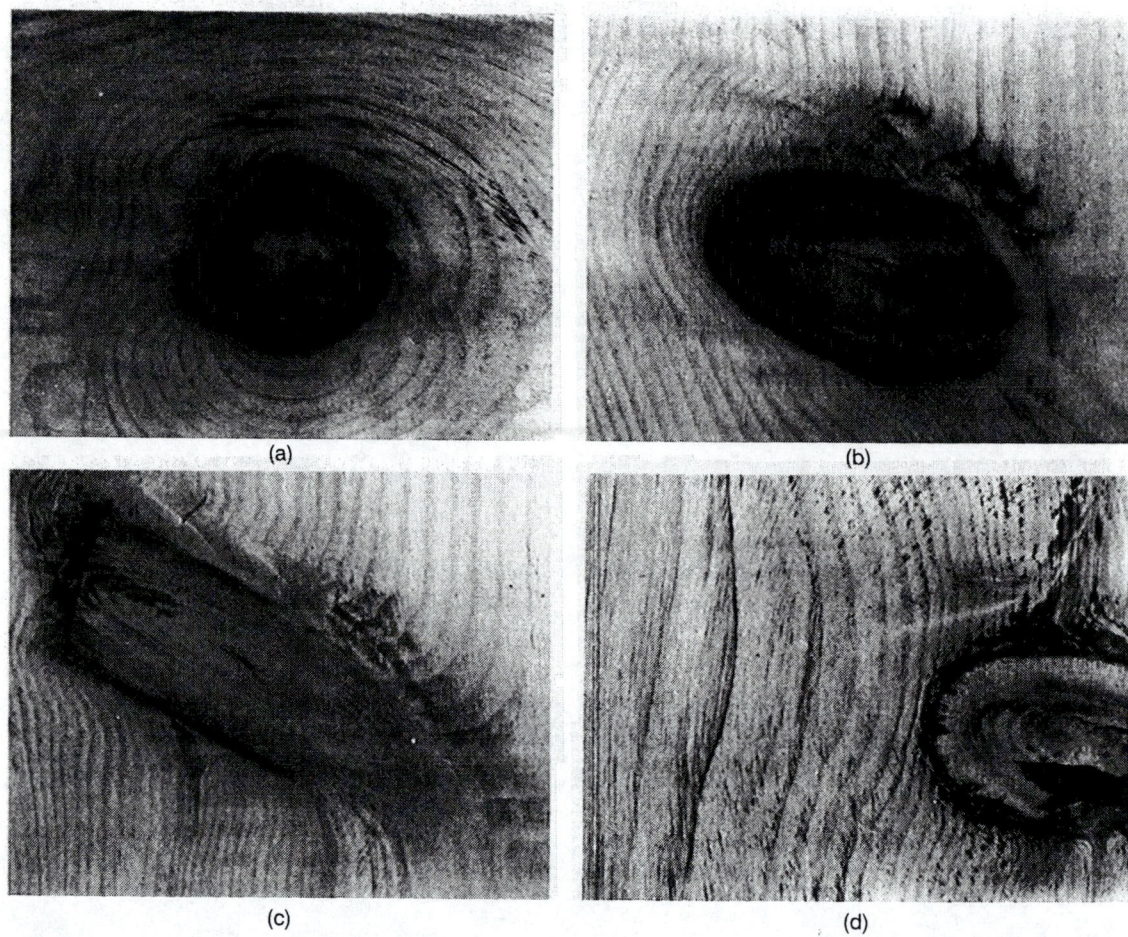

**Figure 10.8**
Some typical knot types: (a) intergrown, (b) oval, (c) spike, (d) edge

affected by the growth structure of the specific wood piece. For example, southern pine is mechanically strong but displays abrupt changes between earlywood and latewood. The lighter density earlywood is not as strong as the latewood; when the growth rings are widely spaced (less than six per inch), the strength properties are inferior and not acceptable for commercial materials.

When lumber is cut, the green moisture will evaporate over time, reaching an equilibrium amount that will vary according to external temperature and humidity conditions. The green moisture fills the fibrous cavities and is partly absorbed in the cell walls. When it is lost, shrinkage takes place. Shrinkage is nearly twice as much in the tangential direction as it is in the radial direction and can lead to **warpage** and twisting of the lumber, as shown in Figure 10.9. When defects such as knots are present, both the degree and direction of the distortion are affected.

**Figure 10.9**
Characteristic shrinkage and distortion of flats, squares, and rounds as related to the direction of annual rings. Shrinkage in the tangential direction is about twice that in the radial.

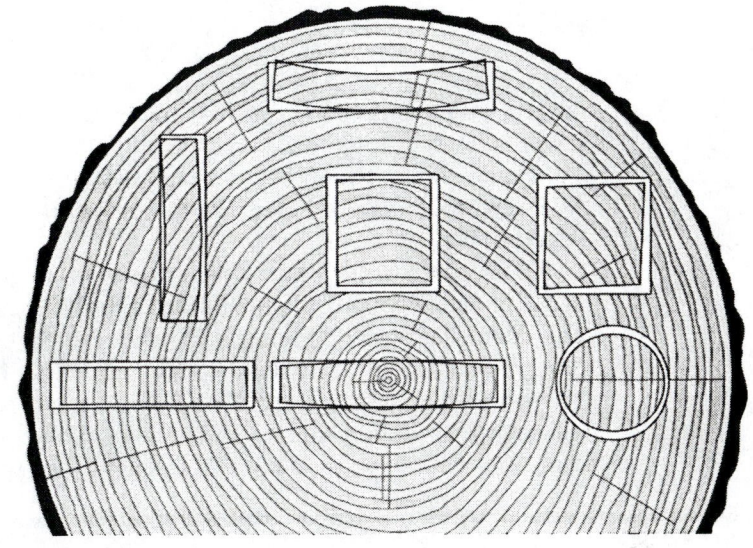

# 10.2.1 Mechanical Properties of Lumber

Lumber used for structural purposes is most commonly softwood, such as white pine, which is cut, dried, and visually inspected or mechanically graded. Both natural seasoning (air drying) and kiln drying are practiced to reduce the moisture content of the lumber to that which is desired for use. Some of the criteria of visual grading are density, slope of the grain, knots, and checks or splits caused by shrinkage stresses in drying. As drying takes place, the wood becomes denser (shrinks) and stronger but because wood is also hygroscopic, this effect on mechanical strength is reversible to some extent. Density is also affected by the growth rate because of the difference in earlywood and latewood. Slope of the grain refers to the anisotropy or directionality caused by the fibrous growth; the slope refers to the angle of the grain only with respect to the longitudinal direction. However, this direction can be any-where from radial to tangential. In some but not all cases, this secondary effect is important. For example, if properties in the radial and tangential direction are the same, then the same properties measured at 45° are about half, whereas when the wood is stronger in the radial direction, the tangential and 45° properties are similar.

The mechanical properties that are reported for wood are different from those reported for steel, for example, because of these directional characteristics and the effect of moisture. In wood, it is difficult to test the tensile strength parallel to the grain and few data exist, so a property called the modulus of rupture is usually reported as a conservative estimate of tensile strength. The **modulus of rupture** (MOR) is given by the outer fiber stress (tensile) measured in three-point bending to failure.

$$MOR = \frac{3P\ell}{2wt^2}$$

where $P$ is the failure load, $l$ is the length between supports, $w$ is the specimen width, and $t$ is the specimen thickness. The modulus of rupture is not a true stress because the formula is not valid beyond the elastic limit, but it is an accepted criterion for the strength of wood. More common properties of hardwoods and softwoods are listed in Table 10.1.

The presence of knots has a major influence on mechanical properties of wood. In fact, wood containing knots almost always fails at loads significantly lower than does clear wood. Although the major effect of knots is that of stress concentration, we must consider grain distortion around a knot, orientation of the knot relative to the wood fiber, different densities and cell structure, and discontinuities (checking) in either knot or surrounding wood. Therefore, in any critical application, such as for planks to be used in staging, the wood must be knot-free.

## Case Study 10.1

### The Broken Stepladder

Falls from ladders are common causes of industrial injuries, and most falls are caused by human error, such as overreaching. This was not so for a maintenance worker who was painting the offices in the Engineering Department at Independent Heat Treat Company, a manufacturer of sintered powder metal and ceramic parts. The worker was painting the top edge of the hallway wall with a brush while standing on the third step of a 6-ft wood stepladder. The hallway was a ramp between two buildings with a minor slope downward to the right as the worker was facing the wall. The fall occurred when the right side rail of the ladder collapsed inward, as shown in Figure 10.10.

Failure analysis showed the break to be caused by bending, but there were no unusually large stresses applied. The materials used in wood stepladders must conform to American National Standards Institute Standard A14.1. This standard specifies that low-density wood shall not be used (low-density wood is defined as exceptionally light in weight and usually deficient in strength properties for the species). Low-density wood is frequently indicated by exceptionally wide growth rings with a low proportion of latewood. From the design viewpoint, the standard specifies minimum side rail thickness of $3/4$ in., which provides for the cutting of a groove $1/8$ in. deep for the supporting step. Thickness must be increased when grooves of greater depth are used.

The side rail of the failed stepladder was $3/4$ in. thick, but the groove where failure occurred was $3/16$ in., 50% deeper than permitted by the standard. Examination of the wood showed it to be a yellow southern pine, acceptable for ladder construction, but there were only five growth rings per inch, with earlywood thickness about twice that of latewood. This is illustrated by the low-magnification scanning electron micrograph in Figure 10.11. Details of the earlywood fracture and latewood fracture appear in Figures 10.12 and 10.13, respectively. In addition, the orientation of growth rings was 45° to the direction of the bending force, the weakest orientation in such wood.

**Table 10.1**
Typical applications and mechanical properties of woods grown in the United States

| Species | Applications | Condition | Specific gravity | Modulus of rupture (psi) | Compression parallel to grain, maximum crushing strength (psi) | Compression perpendicular to grain (fiber stress at elastic limit) (psi) |
|---|---|---|---|---|---|---|
| **Hardwood** | | | | | | |
| Aspen | Lumber | Green | .36 | 5400 | 2500 | 210 |
| (Big tooth) | | Dry | .39 | 9100 | 5300 | 450 |
| Yellow birch | Doors, | Green | .55 | 8300 | 3380 | 430 |
| | cabinets | Dry | .62 | 16,600 | 8170 | 970 |
| Elm | Crates, boxes | Green | .46 | 7200 | 2910 | 360 |
| (American) | | Dry | .50 | 11,800 | 5520 | 690 |
| Sugar maple | Flooring, | Green | .56 | 9400 | 4020 | 640 |
| | furniture | Dry | .63 | 15,800 | 7850 | 1470 |
| White oak | Flooring, | Green | .60 | 8300 | 3550 | 670 |
| | RR ties | Dry | .68 | 15,200 | 7440 | 1070 |
| **Softwood** | | | | | | |
| White cedar | Poles, | Green | .29 | 4200 | 1990 | 230 |
| (northern) | shingles | Dry | .31 | 6500 | 3960 | 310 |
| Red cedar | Poles, | Green | .44 | 7000 | 3570 | 700 |
| (eastern) | shingles | Dry | .47 | 8800 | 6020 | 920 |
| White pine | Lumber, | Green | .34 | 4900 | 2440 | 220 |
| (eastern) | trim | Dry | .35 | 8600 | 4800 | 440 |
| White pine | Lumber, | Green | .35 | 4700 | 2430 | 190 |
| (western) | trim | Dry | .38 | 9700 | 5040 | 470 |
| Pine | Ladder rails | Green | .54 | 8500 | 4320 | 480 |
| (southern | Planks | Dry | .59 | 14,500 | 8470 | 960 |
| longleaf) | | | | | | |

**Figure 10.10**
Broken side rail of wood
stepladder

**Figure 10.11**
SEM micrograph of fractured piece of stepladder side rail showing

**Figure 10.12**
Higher magnification SEM micrograph of earlywood fracture of side rail in
Figure 10.11 (100×)

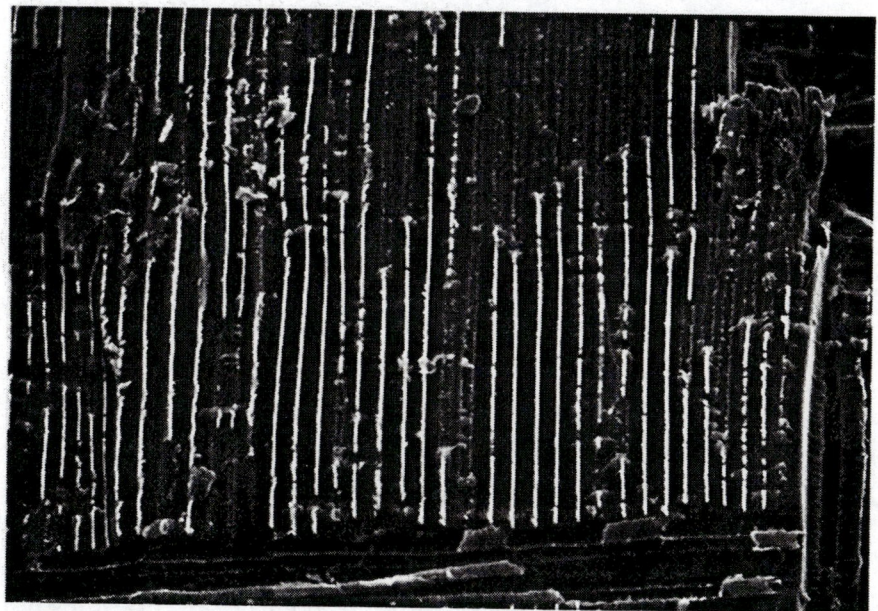

**Figure 10.13**
Higher magnification SEM micrograph of latewood fracture of side rail in
Figure 10.11 (100×)

It was concluded that the failure was caused by the weak side rail and that the worker did nothing to contribute to the accident and injury. This appeased the safety office and prompted more careful inspection of incoming wood stepladders.

## 10.2.2   Electrical, Thermal, and Chemical Properties of Wood

We normally think of wood as an electrical and thermal insulator, but this is true only when the wood has low moisture content. In the live or green condition, wood is conductive; this is also true when wood absorbs moisture. Because of its high electrical resistance when dry, though, wood is used for electric utility poles (protected from absorbing moisture by creosote) and for tool handles for use around electricity. We know that good electrical resistance is synonymous with good thermal resistance and wood is a good thermal insulator.

We also know that wood is combustible—it burns. This good thermal resistance is important in structural applications of wood because wood away from the charred or pyrolized surface remains relatively cool. Thus load-bearing members retain their strength for long periods of time under fire conditions. This means that evacuation can proceed safely without concern of building collapse, at least during the early stages of fire.

Wood is also used for some applications because of its resistance to chemicals. For example, it is used extensively for cooling towers where the boiler waters being cooled have been treated with chemicals and algae-destroying chlorine. Although the chemicals do cause swelling of the wood, the reaction is reversible. There are some reactions that occur, however, that weaken the wood and are not reversible. If steel nails are not galvanized and are used to join wood in the outdoors, for example, exposure can produce iron salts that soften and discolor the wood around the corroded nail.

## 10.3   Modified Wood for Protection Against Decay

Wood is long-lasting only if it is maintained under the proper conditions or if it is **modified**, that is, treated for protection. The major causes of decay are fungi and insects that feed upon the wood under the right conditions of temperature and moisture. Most fungi are inactive below about 50°F and much above 90°F. Simple precautions before using the wood are all that is usually needed to protect it against decay.

**Creosote**, a coal tar derivative, is one of the most useful **preservatives** because it is highly toxic to the organisms that destroy wood, is inexpensive, water-insoluble, and easily applied. However, its use is limited where appearance, paintability, and odor are not important. Workmen also complain because the creosote soils clothing upon contact.

Water-repellant preservatives or stains are dissolved in solvents such as mineral spirits that enable us to apply them easily to wood. The organic chemicals are chlorinated phenols or copper-base organics. Inorganic formulas containing copper, chromium, and arsenic are water-soluble preservatives. The effectiveness of these preservatives depends on the penetration depth and on the retention over time. In many instances, penetration depth is increased by pressure treatment (impregnating under high temperatures and pressures). In some cases, however, preservative solvents are not easily removed after pressure treatment, leaving the wood in an unpaintable condition. Decks made of pressure-treated lumber, for example, cannot be stained until the following season.

# 10.4    Wood Composites

Although wood itself is a composite, there are many ways to make **restructured wood**. It is possible to produce **laminates** such as plywood that have better mechanical properties overall than those of lumber and, at the same time, produce useful sizes and shapes that cannot be made from solid wood. These wood composites also serve to reduce waste because more of a tree is usable as flakes, fibers, and wood flour.

The simplest restructuring is laminating, which reduces the influence of defects because the probability of defects occurring in the same area of two adjacent layers is remote. In addition, thinner layers can be bent to produce large curved sections while reducing warpage and splitting because of the difficulty in drying larger sections. Besides laminating, restructuring of mill products such as trim can be done by cutting end-joints with a large surface area for adhesive joining, thus reducing the amount of scrap from these products significantly.

## 10.4.1    Plywood

**Plywood** is a panel made from layers of veneer with consecutive layers laid at 90° to each other. **Veneer** is a thin layer peeled from a green log, as shown in Figure 10.14. These veneers are bonded together with adhesives, which must be waterproof for exterior grades of plywood. There is only one finished surface, which will be the exposed surface for any application, for example, the finished surface of paneling. Veneers are coated with phenolic resins, full-strength for waterproof properties or thinned for cost savings when waterproof characteristics are not necessary. Coating

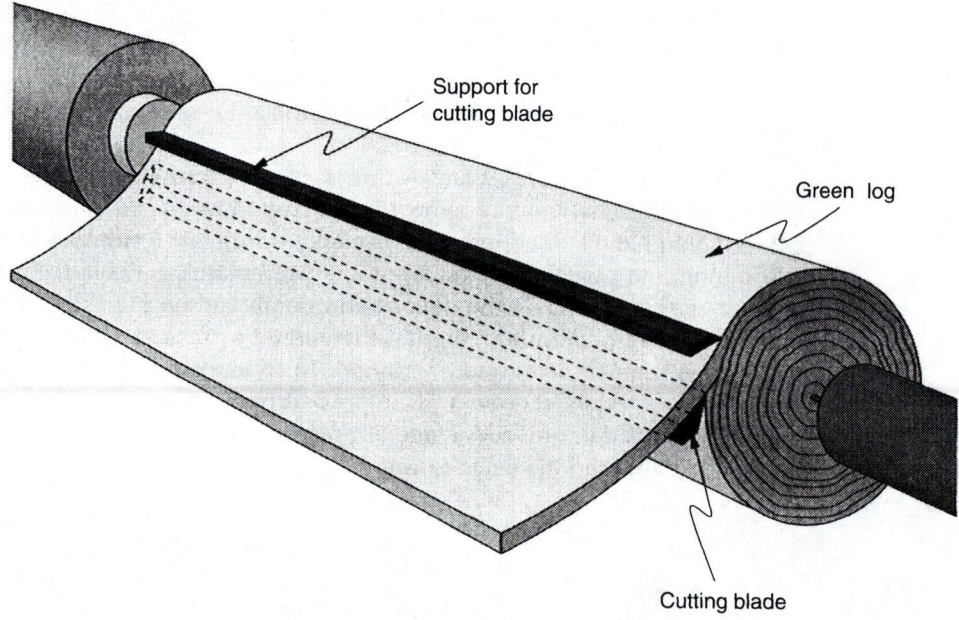

**Figure 10.14**
Schematic diagram for lathe cutting of wood veneer

is usually done by passing the veneer through a rolling mill that transfers the adhesive, stacking it with the proper orientation, and then applying pressure (and heat in some cases) during curing.

There are two classes of plywood—construction, or *industrial*, grade, which is made from softwoods such as southern pine, and decorative, or *hardwood*, grade, such as wall paneling. Most of the former is produced in the United States and most of the latter is imported. In all plywoods, balanced construction is important. Plies should have the same thickness and be the same wood, particularly with respect to the shrinkage, density, and moisture content at the time of bonding. An odd number of plies provides balance, preventing warpage when moisture content changes, but an even number of plies can also be balanced by having the grain parallel in the two center plies. With respect to properties, for construction, or industrial, grade plywood, the modulus of rupture is 7000–10,000 psi and the compressive strength is 4500–6000 psi.

# 10.4.2   Particleboard

**Particleboard** consists of wood particles in the form of chips or flakes (mostly aspen) and finer particles called **furnish** that are bonded with adhesives and wax to form panels or other shaped structures. Thermosetting adhesives that require heat and pressure during curing are normally used. Some typical configurations are shown in Figure 10.15.

(c)

(b)

(a)

**Figure 10.15**
Basic particle panel products: (a) particleboard, (b) waferboard, and (c) oriented strand board

Finer particles, some of which even approach the consistency of wood flour, can be combined with moldable sheet, then shaped under pressure and heat to form products such as toilet seats and Formica countertops. Wood flour is even combined with resin to form "plastic wood" for repair of other wood products.

# 10.5   Paper

**Paper** derives its name from papyrus, the plant from which the ancient Egyptians made the first writing material. True paper, however, has complete separation of the fibers and its invention is credited to Ts'ai Lun of China in the year 105 C.E. Today, paper is still made by essentially the same method originally used in China—pulping to separate the fibers of wood, refining, dilution and straining, pressing, drying, and finishing. Paper is used for many applications, but the main ones are for written or printed documents, for packaging, for absorption, and for structural purposes as a part of composites.

The overwhelming source of fibers in papermaking is wood, although other cellulose fibers from rags (cotton fibers) are used for more expensive papers and in areas of the world where wood is scarce. Both hardwoods and softwoods are used, but for different reasons. Hardwood fibers are shorter and not as strong as softwood fibers; however, they tend to be stiffer, adding bulk, opacity, and smoothness to the paper. Other materials can be added as well, such as clay for opacity and $TiO_2$ or other pigments for color.

In the following sections, we will follow the papermaking processes and examine the structure of various papers as it relates to specific applications.

## 10.5.1  Separation of Wood Fibers

We can separate wood fibers mechanically or with chemicals. Traditionally, mechanical **pulp** has been produced by grinding wood with a stone to produce groundwood, whereby debarked logs, some 4 ft long, are fed into a rotating abrasive stone surface, forming chips that are about $1^1/_2$ in. long and $1/_4$ in. thick. Today, logs are fed into a rotating disc containing a number of cutting blades. Chips are then fed into a rotating disc refiner, where they are broken up into fibers that are progressively refined as they move outward. The most common chemical pulping process is the kraft process whereby chips are digested (or cooked) with sodium hydroxide and sodium sulfite, then separated and thoroughly washed. Screening removes undigested chips and knots. Pulp is bleached, then mechanically refined, or *beat*, removing the primary fiber wall, shortening fibers but increasing their flexibility. Beating also causes unraveling or defibrillating of the fiber walls, as shown in Figure 10.16.

## 10.5.2  Making Paper

Paper is made by a *felting* process, similar to that originated by the Chinese, whereby a dilute suspension of the swollen wood fibers was made into a flat sheet by straining the suspension through a screen. Once the mat was dried, the fibers adhered to each other where they were in contact without needing any other additives. This feature distinguishes paper from other sheet materials such as textiles, which are interwoven for strength. Paper was made by hand until the nineteenth century, when the first continuous machines were developed. Today, paper is produced on **Fourdrinier paper machines** (named after the inventors of the first commercially successful machines), such as the one depicted in Figure 10.17.

The suspended paper stock is introduced uniformly across the machine through the flow spreader, then the headbox distributes it onto the moving wire of the Fourdrinier table where dewatering occurs by vacuum-assisted drainage. In the press section, more water is removed and the paper is consolidated. By pressing the fibers together, larger areas of the bonds are formed at the fiber intersections. After further drying, the sheet is pressed to smooth the surface and make the sheet thick-

**Figure 10.16**
SEM micrograph of refined fibers displaying unraveling of microfibrils (2500×)

ness uniform. Finally, the sheet is wound onto a reel, from which it will be cut or slit into many sizes, stacked or rereeled, and packaged for shipment.

## 10.5.3  Structure of Some Paper Products

There are thousands of different paper products available in the modern marketplace. How are the disparate properties that are needed in a particular product achieved? Facial tissue, for example, must be soft yet absorbent and have some strength. Figure 10.18 shows that the absorbency and softness are derived from very long, loosely bonded fibers, and strength is achieved by using several layers, or plies. These products have to be supported throughout the forming, pressing, and drying processes in order to produce the low-density sheet. For paper toweling, however, strength plus absorbency are critical properties. In these products, support through all procedures is also required, but nonuniform pressing provides the added strength needed for the toweling where it is pressed while areas not pressed are left open to retain absorbency. Figure 10.19 shows the bundling from pressing and the open absorbent areas of a commercial paper towel.

When paper is made for writing or printing, some of the important properties are opacity, permeability, and strength or durability. In newspaper printing, however, strength is secondary to opacity—newsprint is not expected to endure. Figure 10.20 shows the large number of hardwood vessels in newsprint, which provide opac-

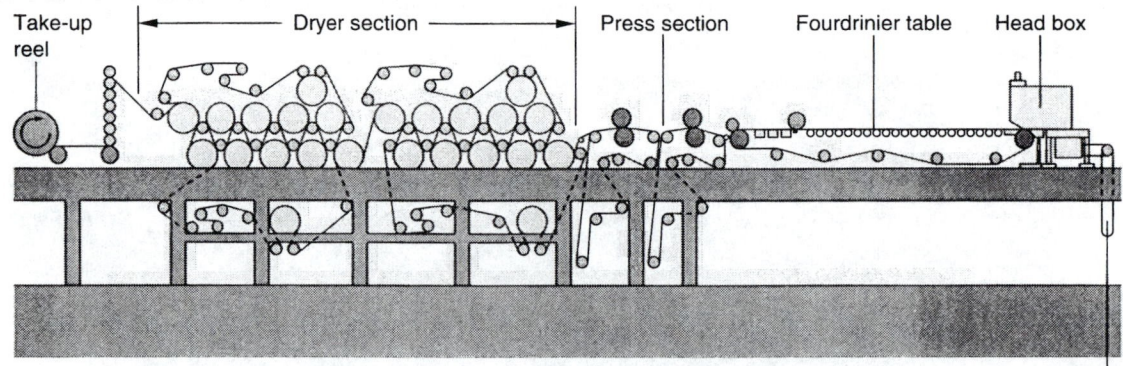

(a)

(b)

**Figure 10.17**
Fourdrinier paper machine: (a) schematic diagram, (b) view of machine in operation
(Fourdrinier table at right and take-up reel at far left)

**Figure 10.18**
SEM micrograph of facial tissue (250×)

**Figure 10.19**
SEM micrograph of absorbent paper toweling (250×)

**Figure 10.20**
SEM micrograph of newsprint (250×)

ity but not strength. On the other hand, a writing pad is expected to last for a long time yet must also be inexpensive. Figure 10.21 shows that these criteria are met with softwood tracheids and some fine particulates added for opacity. When permanent records are desired, as in this textbook, paper surfaces can be coated with clays to provide brightness, opacity, and low permeability. Figure 10.22 shows the surface of a textbook page, where all the fibers are covered with clay containing $TiO_2$, a white pigment. The edge of the page in this figure was torn intentionally and Figure 10.23 shows the exposed fibers and a crack in the brittle clay surface.

## 10.5.4  Packaging

Whenever paper pulp is being distributed onto the Fourdrinier wire, there is a tendency for fibers to align in the machine direction, a tendency that increases for thicker sheets. Figure 10.24 illustrates this anisotropy in an ordinary shopping bag, where the machine direction is top to bottom. Perhaps the directionality is most apparent when we think how easy these bags tear from the top but not from side to side.

For stronger papers (called *linerboards*) used for wrapping, strength is improved by adding other cellulosic fibers, such as jute plant fibers, that are much longer. Figure 10.25 shows how these long fibers become intertwined with many cross fibers when laid onto the Fourdrinier wire. Linerboard is also used as the exterior surface of corrugated boxboard; the inner fluted layer, called *medium*, is used mainly for stiffness.

**Figure 10.21**
SEM micrograph of writing pad (250×)

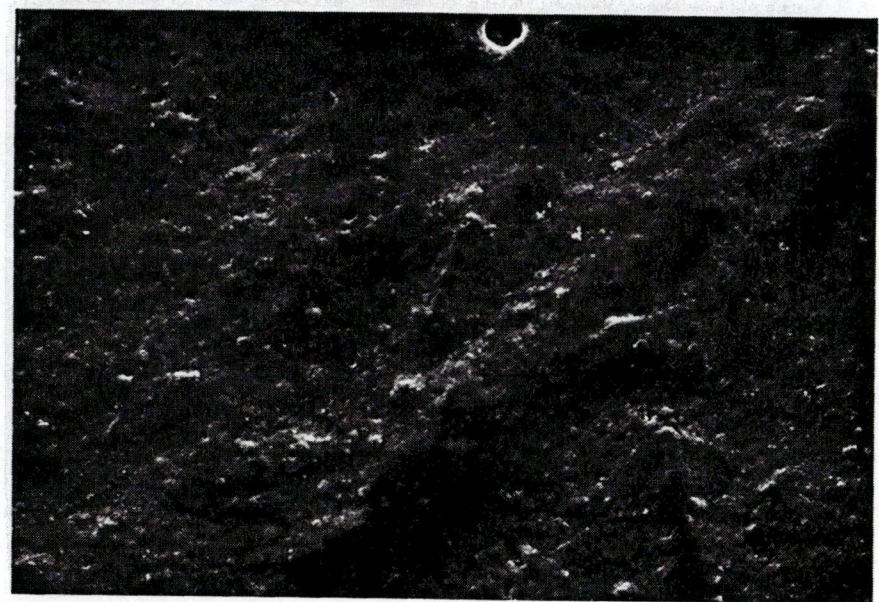

**Figure 10.22**
SEM micrograph of textbook page (250×)

**Figure 10.23**
SEM micrograph of edge of torn textbook page (600×)

**Figure 10.24**
SEM micrograph of fiber alignment in machine direction (top to bottom)
of shopping bag (250×)

**Figure 10.25**
SEM micrograph of boxboard containing long jute fibers (40×)

# Summary

Wood is one of the most widely used industrial materials but occupies a class of its own because it was once a living cellular structure. As a structural material, though, it is classified as a natural composite because it comprises fibrous cellulose and lignin. The growth of trees occurs at the interface of the bark and sapwood, known as the cambial zone, which creates anisotropy in lumber cut from logs. Highest strength is in the fiber direction, but radial and tangential properties are dependent on the section where the log has been cut. Because wood contains moisture, the anisotropy can lead to distortion upon drying as well. Properties are also affected by knots that may be present in different forms, such knots acting as stress concentration factors. Wood is categorized as hardwood, obtained from deciduous trees, and softwood, obtained from conifers. Wood composites such as plywood and particleboard restructure the wood to compensate for directional properties.

Another entire industry, the paper industry, is also based on wood. Wood chips are chemically treated or mechanically ground to separate the fibers, which are rinsed, bleached, refined, diluted in water, and then laid onto wire. The fibers bond together to provide paper of many qualities and uses, from information storage to packaging.

# Terms to Remember

| | |
|---|---|
| anisotropy | modified wood |
| cambium | modulus of rupture (MOR) |
| cellulose | moisture content |
| conifer | paper |
| creosote | particleboard |
| deciduous | plywood |
| earlywood | preservatives |
| fiber | pulp |
| Fourdrinier paper machine | rays |
| furnish | resin canal |
| growth ring | restructured wood |
| hardwood | sapwood |
| heartwood | softwood |
| hygroscopic | tracheid |
| knot | veneer |
| laminate | vessel |
| latewood | warpage |
| libriform | wood |
| lignin | |

# Problems

1. Describe in your own words the differences between softwoods and hardwoods.
2. Relate a personal experience when you had to select and use a wood or wood product for a specific application.
3. Describe the process of photosynthesis, explaining the terms *bark*, *cambium*, *sapwood*, *heartwood*, *lignin*, and *cellulosic fiber* structures in your answer.
4. Explain in your own words how the strength and straightness of a two-by-four depends on its location in the live tree trunk.
5. Describe in your own words how knots affect the strength of structural woods.

6. Explain why visual inspection is so important to a carpenter about to purchase each of the following products:
   a. two-by-fours
   b. a wood stepladder
7. Explain in your own words how plywood is made. What are the differences between industrial grades and hardwood grades?
8. Describe three processes for making useful products out of wood scraps.
9. Describe in your own words how wood is prepared for papermaking.
10. Describe in your own words the structure of the following products:
    a. the writing paper this question is answered on
    b. your textbook
    c. a shipping carton
    d. a paper milk carton

# Glossary

**Abrasion resistance**  Resistance to abrasive wear, closely related to compressive strength
**ABS (acrylonitrile-butadiene-styrene)**  Family of terpolymers that are strong and tough
**Acrylic**  A synthetic resin prepared from acrylic acid
**Additions**  Substances such as plasticizers or colorants that are added to polymers for specific purposes
**Adhesion**  The action or state of substances sticking together
**Admixture**  A material other than cement, water, or aggregate that is added to concrete
**Aggregate**  Inert material such as crushed stone that is combined with portland cement and water to make concrete
**Air classification**  Separation into particle-size fractions by means of an air centrifuge
**Air-entrainment**  Incorporation of billions of tiny discrete air bubbles in portland cement to improve the resistance of the concrete to freeze-thaw conditions
**Aliphatic**  Organic molecules that have no rigid benzene rings in their structure
**Alkyd**  Polyester resins modified with fatty acids, used in oil-base paints
**Amorphous**  A noncrystalline solid lacking long-range atomic order
**Anisotropy**  Display of different properties when measured along different directions
**Annealing point**  Temperature at which complete annealing is accomplished
**Aromatic**  Hydrocarbon molecules characterized by presence of benzene ring structures

**Atactic**   A molecule in which side groups are randomly arranged along the polymer chain

**Atomic packing factor**   The fraction of a unit cell occupied by atoms

**Bakelite**   Trade name for phenol formaldehyde discovered by L. H. Baekeland

**Blast furnace slag**   A cementitious material sometimes added to concrete

**Block copolymer**   A polymer molecule that is made up of long sections of one composition separated from segments of a different composition

**Blow molding**   Fabrication method whereby containers are made by forcing a parison against a shaped mold by internal pressure

**Blown films**   Film produced by extrusion over a mandrel, then expanded by air pressure applied through the mandrel

**Borosilicate**   Glass containing boron oxide, $B_2O_3$, as part of network

**Brittle**   The behavior of a material characterized by little or no plastic deformation prior to fracture

**Calcium hydroxide**   A crystalline material formed by hydration of portland cement

**Calcium silicate hydrate (C-S-H)**   Tobermorite gel formed by hydration of portland cement that is responsible for most of the strength of concrete

**Calendering**   Process for blending and for making continuous sheet by passing material through sets of rolls to control thickness and finish texture

**Cambium**   A thin layer of cells between the bark and wood of a tree that continually subdivides to form new wood and bark cells

**Carbon-matrix carbon-fiber composites**   High-temperature stable composites developed from resin matrix precursors and carbon fiber in preform configurations

**Cast films**   Product of dissolution of thermoplastic polymers with plastisols, then sheet casting to form films upon evaporation of the plastisols

**Casting**   Process for pouring fluid polymers into a shaped mold where they take the shape of the mold as a solid after cooling or polymerization

**Casting processes**   Methods for casting polymers or ceramics

**Cellulose**   The carbohydrate that is the major constituent of wood and forms the framework for all wood cells

**Cementitious**   Having the bonding properties or behavior of portland cement

**Centrifugal casting**   Casting a polymer into a mold that is rotated on one or more axes while the polymer cools or polymerizes to take the shape of the mold

**Ceramic matrix composites**   Composites that have reinforcement of the ceramic phase

**Clay**   Naturally occurring mineral of hydrated aluminosilicate of fine particle size

**Coatings**   Permanent layers deposited on a substrate

**Collimated**   Parallel alignment in the axial direction

**Composition**   The chemical analysis of a material

**Compression**   The squeezing action caused by forces directed at each other

**Compression molding**   Process whereby a polymer is placed into a die cavity, then heat and pressure is applied to cure the polymer

**Compressive strength**   The highest engineering compressive stress that a material can withstand without fracturing

**Concrete**   The composite material based upon aggregates bonded together by hydrated cement

**Condensation**   Polymerization process whereby a chemical reaction forms the polymer and a simple by-product

**Conductivity**   The ability of a material to carry electrical current; the reciprocal of electrical resistivity

**Conifer**    Group of trees that in most cases have needlelike or scalelike leaves

**Constituents**    The ingredients of a material

**Continuous**    Characterized by no interruption in space or length

**Coordination number**    The number of nearest neighbor atoms that an atom has in a specific crystalline or noncrystalline structure

**Copolymer**    An addition polymer that contains two types of mers

**Covalent bond**    The directional atomic bond created when valence electrons are shared by two adjacent atoms

**Creosote**    A coal tar derivative used for preservation of wood

**Crystallinity**    The structure of some polymers where the molecular chains are uniform and compact, forming solid crystals that have definite geometric form

**Crystal structure**    The long-range geometric array of atoms bonded together in a solid

**Curing**    The process of hydration of concrete or polymerization of thermosetting resins

**Deciduous**    Adjective used to describe trees that lose their leaves at the end of the growing season

**Denier**    The mass in grams of 9000 meters of synthetic fiber; translates to filament diameter when density is known

**Densification**    The elimination of porosity that takes place in sintering

**Density**    The ratio of weight to volume of a material

**Design joints**    Joints in concrete designed to prevent cracking or to control cracking for cosmetic purposes

**Diamond**    The hardest substance known, attributable to the tetrahedrally coordinated crystal structure of the carbon atoms

**Dicalcium silicate ($C_2S$)**    An ingredient of portland cement that reacts during hydration to provide bonding of aggregates

**Dielectric strength**    The measure of the highest resistance of a material to an applied electric current before breakdown (conduction) occurs

**Dip casting**    The process of immersing a mold into a resin; after cooling, the part has the internal shape of the mold

**Discontinuous**    Characterized by interrupted length or space; in composites, discontinuous means short fiber lengths

**Dispersion strengthened**    Enhanced strength obtained by fine spherical powders embedded in a weak matrix

**Doctor blade**    A device used to form thin ceramic sheet preforms for utilization as a substrate

**Drape molding**    Method for shaping thermoplastic materials whereby the polymer sheet is heated, then draped over a mold and drawn to it by vacuum

**Dry pressing**    Consolidation of particles under pressure with no binder present

**Ductility**    The ability to deform without fracture, usually reported as percent elongation or percent reduction in area

**Durability**    Resistance to degradation of any type over time

**Earlywood**    The portion of a tree's annual growth ring formed during the early part of the growing season, usually less dense and weaker than latewood

**Ejection**    Removal, usually by pins, after injection molding

**Elastic deformation**    Deformation of a material under stress that is recovered when the stress is removed

**Elastic modulus**    The ratio of stress to strain during elastic deformation

**Elastomer**    A polymer that can be stretched elastically 200% or more

**Electrical properties**   Material properties such as conductivity or resistivity in the presence of an electric field

**Elongation**   The change in length caused by an external tensile stress, used as a measure of ductility

**Encapsulation**   Enclosing an item, such as an electronic component, in plastic for protection

**Endurance limit**   The maximum strength of a material subjected to fatigue stresses

**Epoxy**   A thermosetting polymer based upon ethylene oxide

**Ettringite**   A needle-shaped trisulfate formed in the early stages of curing of concrete

**Expandable styrene**   A two-stage product with pre-expanded beads placed in a mold and expanded further to shape

**Expanded foam**   Plastics that are expanded in a single-stage process to produce a spongelike or cellular product

**Extrusion**   A direct compression process whereby material is shaped by squeezing it through a die

**Fabrication**   The manufacture of material into useful shapes

**Fabric ply**   One preimpregnated sheet of a laminate

**Fatigue**   The intermittent or cyclic application of stress to a material

**Ferrimagnetism**   A type of magnetism that depends on a net magnetization resulting from unequal atomic magnetic dipoles aligned in opposite directions

**Fiber**   A long slender material used to make both woven and unwoven materials, or as a strengthener for composite materials

**Fiber reinforcement**   Strengthening by addition of strong fibers to a weak matrix

**Filament winding**   A process for making fiber-reinforced polymers whereby the fibers are wound onto a rotating mandrel, then impregnated with resin and cured

**Fillers**   Additives to A-stage resins usually made to improve properties or reduce cost

**Films**   Thin coatings of polymers applied to other materials

**Flux**   A low-melting constituent, such as in a ternary crystalline ceramic

**Fly ash**   An admixture for concrete that has cementitious and pozzolanic properties

**Fourdrinier paper machine**   A high-speed machine that deposits processed fibers on moving wire where drying occurs; the paper sheet is then wound onto a roll for further processing

**Fracture toughness, $K_{1c}$**   The critical value of the stress intensity factor necessary to propagate a crack to complete failure by fracture

**Freeze-thaw resistance**   The resistance to cracking of concrete caused by the expansion of water when it freezes

**Furnish**   Wood material that has been reduced for use in wood-base fiber or particleboard products

**Fused silica**   A high-temperature crystalline glass

**Glass former**   A compound that forms a glass network

**Glass modifier**   A compound that does not join a glass network but changes it

**Glass network**   The noncrystalline three-dimensional structure of glass

**Graft copolymer**   A copolymer that has one of the constituent mers attached to the second mer and acts as the backbone of the structure

**Graphite**   The layerlike crystalline form of carbon

**Growth ring**   The layer of wood growth put on a tree during a single growing season

**Gutta-percha**   A natural latex that was the first commercial polymer material

**Gypsum**   A calcium sulfate added to portland cement to promote early-stage curing

**Hardness**   Resistance of a material to permanent indentation by a shaped indenter under load

**Hardwood**   Deciduous trees that shed their leaves at the end of the growing season

**Heartwood**   The wood that extends from the center of a tree to the sapwood and no longer takes part in the life process of the tree

**High performance**   Achievement of maximum properties of a material

**Hydration**   The chemical reaction of portland cement with water that produces a solid material that bonds aggregates together in concrete

**Hydraulic cement**   A cement such as portland cement that hydrates to form a solid that resists water

**Hygroscopic**   Adjective meaning moisture absorptive

**Impact**   A method for measurement of toughness of a material whereby the specimen is struck with a kinetic force

**Injection molding**   A fabrication method whereby heat-softened plastic is injected into a mold, cooled, and ejected

**Inorganic**   Adjective meaning not organic, that is, not based upon hydrocarbon molecules

**Ionic bond**   A bond between ions of elements based upon electrostatic attraction

**Isostatic pressing**   Application of pressure uniformly in all directions

**Isotactic**   A molecule in which side groups are always on the same side of the chain

**Kevlar**   An aromatic polyamide that is used as high strength filaments in composites and for bullet resistant fabrics

**Knot**   The portion of a limb or branch that has been surrounded by subsequent growth of the tree trunk

**Laminate**   A product made by bonding together two or more layers of material

**Laminating**   The process of bonding together two or more layers of material

**Laser**   A beam of coherent light radiation

**Latewood**   The portion of a tree's annual growth ring formed during the late part of the growing season, usually more dense and stronger than earlywood

**Lay-up process**   The process of placing successive layers of reinforcing fabric in a shape that is impregnated with resin to form a reinforced composite

**Libriform**   Thick-walled, narrow fibers of hardwoods

**Lignin**   A natural polymer that bonds the individual wood cells together

**Lost-core injection molding**   Injection molding plastic around a core that then is melted away to provide a product with a cavity the shape of the core

**Macroscopic combinations**   Material combinations where interfaces are relatively large

**Matrix**   The continuous phase in a composite structure

**Mechanical properties**   The properties based upon mechanical deformation of a material

**Melamine formaldehyde**   Resin formed by reaction of melamine and formaldehyde that is used in surface coatings, decorative plastics, and dinnerware

**Melt flow rate**   A measure of the amount of plastic (exclusive of polyethylene) extruded in a specific time through a specific small orifice at a pressure of 43.5 psi at a specific temperature above the melting or softening point

**Melt index**   A measure of the amount of polyethylene extruded in a specific time through a specific small orifice at a pressure of 43.5 psi at a specific temperature above the softening point

**Melting**   The process of changing a solid phase into a liquid phase

**Melting temperature**   The temperature at which melting of a material takes place

**Mer**   The smallest repetitive unit of a polymer

**Metal matrix composites**   Composite materials that have a matrix of metal reinforced with a second material

**Microcracks**   Material flaws of microscopic dimensions

**Microstructure**   The distribution and size of phases of a material as seen in a microscope

**Mixing**    Process of distributing different materials uniformly

**Modified wood**    Wood processed by chemical treatment, pressure, or other means to alter the properties substantially from the original wood

**Modulus of rupture**    An accepted measure of the strength of wood in bending

**Moisture content**    Weight of the moisture contained in wood compared to the kiln-dried weight

**Monolithic**    Single-phase macroscopic material

**Monomer**    A simple molecule capable of reacting with other like or unlike molecules to form a long-chain, high molecular weight polymer

**Monosulfate**    The crystals formed by dissolution of ettringite during later stages of the curing of concrete

**Mortar**    The mixture of portland cement with fine aggregates used to bond bricks

**Multidirectional**    Distribution of a phase in many different directions

**NBR**    A nitrile rubber that is the copolymer of acrylonitrile and butadiene

**Neoprene**    A fuel- and weather-resistant rubber similar to natural rubber except that a chlorine atom replaces $CH_3$ in the structure

**Nonmetallic**    Any material that does not have metallic characteristics

**Nylon**    A polyamide used for spun fibers and strong, heat-resistant molded products

**Opaque**    Does not transmit light

**Optical properties**    Properties that describe the response of a material to light radiation

**Orlon**    An acrylic fiber (trademark of DuPont)

**Paint**    A polymer containing pigments, used for coating many materials; may be either water-base (latex) or oil-base (alkyd)

**PAN (polyacrylonitrile)**    A fiber used for conversion to strong graphite fibers for use in composites

**Paper**    The sheet material formed from treated wood fibers

**Particleboard**    A thicker, more rigid grade of paper (typically greater than 0.012 in. thick)

**Particle size**    The size, usually given as the diameter, of particulate materials

**Permeability**    The ability of a substance, usually air or water, to pass through a material

**Perovskite**    The name given to the crystal structure of $CaF_2$

**PET (polyethylene terephthalate)**    A versatile polyester available as film, sheet, or fiber

**Phase**    Chemically homogeneous portion of microstructure

**Phase diagram**    The temperature-composition diagram that describes the relationships of phases at equilibrium

**Phenolic**    A thermosetting polymer based upon the reaction of phenol and formaldehyde (*see* Bakelite)

**Physical properties**    Properties of material not reflecting a mechanical or chemical change

**Plastic**    Adjective referring to pliability for shaping under pressure

**Plasticate**    To melt and mix a thermoplastic polymer

**Plastic deformation**    Change in shape of a solid material caused by an external force

**Plastic forming**    The shaping of materials by deformation processing

**Plasticity**    Capability of being deformed under pressure

**Plasticizer**    An additive for polymers that promotes plasticity

**Plastic matrix composite**    A composite with reinforcement of the polymeric matrix

**Plastisol**    A polymer solvent

**Plywood**    A glued wood laminate made up of thin veneer layers with the grain of adjacent layers at right angles

**Polyacetal**    Common name for polyoxymethylene, used because acetal refers to the oxygen atoms joined to the mers

**Polyamide**    A linear thermoplastic polymer with the -CONH- connecting link

**Polycarbonate**    A linear, amorphous polyester that has high impact strength

**Polyester**    A condensation polymer in two forms—saturated thermoplastics made into fibers and unsaturated thermosetting resins used in many composites

**Polyethylene**    A polyolefin based upon the ethylene monomer $C_2H_4$

**Polymer**    A compound with high molecular weight whose structure can be defined by repeating mers

**Polymer concrete**    Concrete that has a polymer added to or in replacement of the portland cement

**Polymethyl methacrylate**    An atactic, amorphous, transparent polymer used for many optical purposes

**Polyolefin**    A polymer based upon a number of monomers that have linear carbon-to-carbon double bonds

**Polypropylene**    A polyolefin with the vinyl linkage $CH_3$ side group

**Polystyrene**    A polyolefin with vinyl linkage and the benzene ring as the side group

**Polyurethane**    A versatile polymer available as coatings, molding compounds, foams, and more; the recurring link of the polyurethane chain is -NHCOO- or -NHCO-

**Polyvinyl chloride (PVC)**    A polyolefin with vinyl linkage and the chlorine atom as the side group

**Porosity**    The void space in a material

**Portland cement**    The mixture containing calcium silicates, calcium aluminate, and calcium aluminoferrite that hydrates with water and binds aggregates in making concrete

**Pozzolan**    A material that in itself is not cementitious, but when finely ground and reacted with calcium hydroxide forms cementitious compounds

**Precursor**    An original material that undergoes a transformation to another, more useful, material during processing

**Prepregging**    Process for adding resin to a fibrous material prior to partial curing of the resin

**Preservative**    Any substance added to wood to prevent deterioration of the wood caused by fungi or insects

**Prestressed concrete**    Reinforced concrete that has been cured while rebars were under tension and, after curing, tension has been released, compressing the cured concrete

**Processing**    The formation of shaped materials

**Pulp**    The cellulose fibers of wood processed mechanically or chemically for manufacture of paper or paper products

**Pultrusion**    The process for pulling rovings through a thermosetting material, then through a long heated die where polymerization occurs

**Raw materials**    The starting materials selected for processing

**Rays**    Wood cells and structures that extend radially and horizontally from the cambium

**Rebars**    The common term for steel rods used to reinforce concrete

**Reinforced concrete**    Concrete that has been reinforced with steel rebars

**Reinforcement**    A phase added to a matrix for the purpose of strengthening

**Resin**    A viscous fluid obtained from certain plants or made from synthetic materials

**Resin canal**    Intercellular passages in softwood that contain and transmit resinous material

**Resin transfer molding**    A process whereby catalyzed resin is transferred into a mold containing reinforcement

**Resistance**    The electrical property that is dependent on shape and material resistivity

**Resistivity**    The characteristic property of a material that describes the flow of electricity in the specific material

**Restructured wood**   Wood that has been modified by cutting or chemical processing, then restructured and bonded to provide useful products such as plywood, particleboard, and plastic wood

**Rolling**   The deformation or shaping of a material when passed through pressure rolls

**Rotational casting**   A process for making hollow polymer parts whereby the mold is rotated in one or more planes

**Rubber**   A natural or synthetic polyisoprene resin used as an elastomer or matrix for composites

**Rule of mixtures**   The properties of a mixture as influenced by the volume fraction of phases

**SAN (styrene acrylonitrile)**   A transparent copolymer resistant to many solvents and fats

**Sapwood**   The wood nearest the cambium where the life process of the tree takes place

**SBR (styrene-butadiene rubber)**   Synthetic rubber based upon the styrene-butadiene copolymer

**Screw spinneret**   A type of extrusion die that produces fine fibers and filaments that are spun into thread at the exit from the die

**Sheet molding**   Forming of plastic-fiber composites into a sheet

**Shrinkage**   The volume change associated with cooling to a rigid (solid) state

**Silicone**   A polymer based upon silicon and oxygen in the chain instead of carbon

**Sintering**   The process of densification by pore removal of compacted materials at elevated temperatures

**Slump**   The measure of formability of fresh concrete

**Softwood**   The group of trees that have needlelike or scalelike leaves

**Spandex**   A block copolymer with rigid polyurethane and flexible polyether groups

**Strain**   The change in length or cross section with respect to original dimensions caused by a load

**Strain point**   The temperature of a glass below which viscous flow does not occur

**Stress**   The load per unit area applied to a shape

**Stress concentration**   The intensification of any applied stress relatable to geometric factors, usually related to crack propagation and failure

**Superplasticizer**   An additive for polymers that promotes extreme plastic behavior

**Syndiotactic**   A molecule in which side groups alternate regularly on opposite sides of the chain

**Tape**   A fiber-reinforced material in the shape of a continuous ribbon, used to form larger composites

**Technical ceramics**   Modern ceramics that are important in industrial applications

**Temperature gradient**   The temperature variation through a thickness of material

**Tempered glass**   Glass that has been heat treated to place the outer surfaces in compression

**Tensile strength**   The highest engineering tensile stress that a material can withstand without fracturing

**Tension**   The stress caused by application of a force pulling a material apart

**Terpolymer**   A long-chain polymer based upon three separate mers joined in the chain

**Tetracalcium aluminoferrite ($C_4AF$)**   An ingredient of portland cement

**Thermal conductivity**   The property of a material describing flow of heat through the material

**Thermal expansion**   The increase in volume that most materials undergo when heated, usually expressed in terms of linear dimension

**Thermal shock**   The stress caused by a thermal gradient across a material

**Thermoforming** A process of heating thermoplastic sheet materials and forming them over a die

**Thermoplastic** A polymer that can be reshaped by heating and applying pressure

**Thermosetting** A polymer that is cross-linked and cannot be reshaped except by machining

**Three-directional orthogonal reinforcement** Strengthening of a material by alignment of fibers in three orthogonal directions

**Tooling** The assembly of jigs, fixtures, and so on that are used in molding or fabricating parts

**Toughness** The combination of strength and ductility, usually measured by impact testing

**Tracheid** The elongated cellulosic cells that make up the major part of the structure of softwoods

**Traditional ceramics** Ceramics based upon molding and firing of clay, for example, the pottery and brick industries

**Transfer molding** A process whereby resin is heated in a chamber, then transferred to a shaped mold where high pressure completes the shaping process

**Transition temperature** The temperature at which a specific material property changes, for example, the melting point

**Transition zone** The interface between aggregate and cement where bonding takes place in concrete

**Translucent** Transmitting light but scattering it so that images are not discernable

**Transparency** Transmission of light without distortion

**Tricalcium aluminate (C₃A)** An ingredient of portland cement

**Tricalcium silicate (C₃S)** An ingredient of portland cement

**Ultimate tensile strength** The maximum strength exhibited by a material under stress before fracture occurs

**Unidirectional** Responsive in a single direction

**Veneer** A thin layer or sheet of wood

**Vessel** A long tubular structure formed by hardwood cells that have been combined

**Viscoelasticity** The combination of viscosity and elasticity in the same material

**Viscosity** The resistance to material flow

**Vitreous** A synonym for noncrystalline or amorphous

**Vulcanizing** The cross-linking of the polymer chains of rubber

**Warpage** The bending of a material during drying (e.g., wood) or curing (e.g., polymer)

**Water/cement ratio** The ratio of water to portland cement in a concrete mixture; cured strength is affected by this value

**Water-reducing** An adjective describing an admixture for concrete that lowers the water content

**Whiskers** Fine fibers that have large aspect ratios (length to diameter)

**Wires** Large-diameter fibers, usually metal

**Wood** A natural composite material obtained from trees

**Workability** The ease of placing, consolidating, and finishing of concrete before hydration occurs

**Working range** The temperature range where deformation can take place in processing

**Yield strength** The highest stress a material can withstand without permanent deformation

# *Index*

Abrasive products, 104
ABS (acrylonitrile butadiene styrene), 21
Accelerators, 24
Addition polymerization, 19
Admixtures (concrete), 175–177
Aggregates, 164, 165, 170
Air classification, 107, 108
Air-entrained concrete, 174
Alabaster, 97
Aliphatic polymers, 18
Alkyds, 43
Aluminosilicate glass, 97
Amorphous, 91, 93
Anisotropy, 192, 193
Annealing, 98
Annealing point, 99
Aromatic polymers, 18
Atomic packing factor, 73

Bakelite, 18
Banbury mixer, 56
Barium titanate, 83
Binders, 113
Blast furnace slag, 176
Blow molding, 56, 99, 100
Blown film, 50
Boron-aluminum composites, 137

Borosilicate glass, 96, 97
Brittleness, 4, 5, 74, 77

Calcium fluorite, 81
Calcium hydroxide (CH), 166
Calcium silicate hydrates (C-S-H), 166, 167
Calendering, 56, 57
Cambium, 189
Carbon, 84
Carbon fibers, 129
Carbon matrix-carbon fiber composite, 152
Case studies
    A cold weather concreting problem, 179
    Ceramic aluminum oxide gas bearings
        for aerospace vehicles, 107
    Ceramics for advanced turbine engines, 77
    Change the material, change the vendor,
        but change both?, 52
    Corrosion failure, 168
    Cutting tools, 75
    Failure of a polyvinyl chloride pipe, 30
    Fused silica aerospace vehicle windows, 96
    Glass provides an environmentally sound
        solution to a deburring problem, 100
    It's fashionable to recycle, 59
    More corrosion problems, 183
    Polycarbonate processing, 55

Case studies, *continued*
  Selecting a composite abrasive finishing wheel, 135
  Solid propellant rocket motor nozzle development, 141
  Stress concentration again!, 27
  Stress concentration and fatigue, 31
  The broken stepladder, 200
  Today's phenolics, 68
Casting, 67, 110
Catalysts, 24
Celluloid, 18
Cellulose, 189
Cement, 165, 166
Cemented carbides, 125, 127
Centrifugal casting, 68, 99
Ceramic magnets, 87
Ceramic matrix composites, 141, 160
Ceramic processing, 103ff
Charpy and Izod impact test, 9
Clay, 71, 84, 103, 105
Clay products, 104
Colored glass, 97
Commercial glass, 96
Composites, 123ff
Compression molding, 63, 148
Compression wood, 193
Concrete design joint, 184
Concrete mixtures, 168
Concrete strength, 170
Condensation polymerization, 19
Conifers, 189
Continuous fiber-metal matrix composite, 154
Coordination number, 79
Copolymer, 21
Corrosion, 168, 181
Coupling agents, 24
Covalent bond, 18, 84
Cracking of concrete, 184
Creosote, 205
Crystal structure, 78
Curing (hydration), 165, 177

Defibrillation, 208, 209
Denier, 59
Densification, 104, 119
Densities of ceramics, 72

Devitrification, 94
Diamond, 84, 86
Dicalcium silicate ($C_2S$), 165
Dip casting, 68
Discontinuous fiber composites, 133, 139, 158
Dispersion strengthening, 123, 124
Doctor blade, 112
Drape molding, 53
Dry pressing, 113

Earlywood, 193
Ejection pins, 63, 64
Elastic (Young's) modulus, 6
Elastomers, 38
Electrical properties, 13
Enamels, 91
Encapsulation, 67
Engineering strain, 4
Engineering stress, 4
Epoxies, 37
Ettringite, 166
Expandable styrene, 60
Expanded foam, 60
Expansion coefficient of ceramics, 73, 74
Extrusion, 49, 116, 118, 119

Fabric plies, 133, 135
Fatigue, 11
  in fiber composites, 134
Feldspar, 72, 103–105
Ferrimagnetism, 87
Ferrites, 87
Fiberglass, 129
Fiber materials, 127, 128
Fiber-reinforcement, 124, 126
Fiber-reinforcement of concrete, 182
Filament winding, 150
Film casting, 55
Float glass, 99
Fly ash, 176, 177
Flux, 104
Foamed plastics, 59–63
Foaming agents, 24
Fourdrinier paper machine, 208, 210
Fracture toughness, 9
Freeze-thaw resistance, 174
Fulcher relation, 98
Fused silica, 96

Glass, 91ff
  composition, 95, 96
  fibers, 128, 129
  formation, 93
  formers, 93, 94
  forming, 99
  modifiers, 94
  network, 94
  structure, 92
  transition temperature, $T_g$, 93
  viscosity, 98
Glossary, 219
Graft copolymer, 21
Graphite, 84–86
Graphite-epoxy tape, 147
Growth rings, 190
Gutta-percha, 18
Gypsum, 165

Hand throwing, 105
Hardness, 8
Hardwood, 189
Heartwood, 189
Hot isostatic pressing, 157, 159
Hydration (curing), 165, 167

Ionic bond, 78
Impurity control, 107, 108
Inhibitors, 24
Injection molding
  ceramics, 118
  polymers, 51, 52
Isostatic pressing, 113, 116

Jiggering, 105

Kevlar, 34, 130
Knots, 193, 198

Laminations, 56, 59, 205
Latewood, 193
Lay-up processes, 145
Leaded glass, 97
Libriform, 190, 192
Lignin, 189
Liquid phase sintering, 121
Lucalox alumina, 109
Lumber, 199

Mechanical strength of ceramics, 74
Melamine formaldehyde, 36
Melt flow rate, 48
Melt index, 48
Melt infiltration, 160
Melting, 92
Metal matrix composites, 136, 137, 139, 154
Modern ceramics, 106
Modified wood, 204
Modulus of rupture (MOR), 6, 199
Moisture content of wood, 190
Monosulfate, 166, 167
Multidirectional, 133, 135

NaCl structure, 80, 81
NBR (nitrile rubber), 39, 40
Neoprene, 39

Opal glass, 97
Opaque glass, 97
Optical properties, 14, 15

Packaging, 212
Paint, 43
PAN (polyacrylonitrile), 34, 35, 129
Paper, 207
Parison, 56
Particleboard, 206, 207
Particle-reinforced composites, 125
Pauling's rules, 79
Perovskite structure, 83
PET (polyethylene terephthalate), 33, 59
Phase diagrams, 104, 105
Phenolics, 36, 68
Physical properties, 2
Pith, 190
Plaster-of-paris, 110
Plastic forming, 116
Plasticity, 104
Plasticizers, 24
Plastic matrix composites, 130, 145
Plastisol, 67
Plywood, 205
PMMA (polymethyl methacrylate), 34
Polyacetals, 30
Polyamides, 34
Polycarbonates, 35, 55
Polyesters, 33

Polyethylene, 18, 29
Polymer additives, 22
Polymer concrete, 182
Polymer crystallinity, 22, 23, 48
Polymer films, 26
Polymer structure, 20
Polypropylene, 19, 29
Polystyrene, 19, 29
Polyurethane, 43
Polyvinyl chloride, 19, 30
Porcelain, 105
Portland cement, 165
Potter's wheel, 105
Pozzolan, 176
Precursors, 129
Prepreg, 63, 146
Prepreg machine, 146
Prepreg tapes, 145, 146
Preservatives, 205
Prestressed concrete, 181
Processing metal matrix composites, 154
Properties of ceramics, 72
Pultrusion, 64, 66, 68

Quartz, 72

Radius ratio, 79
Rays, 190
Reaction wood, 193
Refractories, 104
Reinforced concrete, 179
Reinforcement, 123
Resin canals, 193
Resin transfer molding, 151, 152
Restructured wood, 205
Rotational casting, 68
Rubber, 38
Rubber mold, 114, 116
Rule of mixtures, 125

Sample problems
   Concrete batch weight, 170
   Elastic modulus for plastic and ceramic, 7
   Glass fiber reinforced composite, 131
   Modulus of rupture for wood, 8
   Thermal stress, 13, 77
   Threaded plastic pipe, 10
SAN (styrene acrylonitrile), 29
Sapwood, 189

SBR (styrene butadiene rubber), 39
Screw spinneret, 50, 51
Sheet molding, 148, 149
Shrinkage, 103, 121
Silica, 72, 103
Silica fume, 176, 177
Silicon carbide fiber reinforced titanium, 159
Silicon carbide whisker-reinforced metals, 160, 161
Silicone, 39, 41
Sintering, 119
SI units, 2
Slip casting, 105, 110
Slump, 169
Soda-lime glass, 96
Softwood, 189
Solidification, 93
Spandex, 44
Spark plug insulators, 115, 117
Specific volume, 92
Spinel structure, 87
Splat cooling, 93
Strain, 4
Strain point, 98
Stress, 4
Stress concentration, 10, 27, 31
Superconductor, 139, 140
Superplasticizers, 175
Surfactants, 175

Tape casting, 112
Technical ceramics, 71, 72, 81, 106
Tempered glass, 101
Tensile properties, 3
Tension wood, 193
Terpolymer, 21
Tetracalcium aluminoferrite, $C_4AF$, 165
Thermal conductivity, 73, 75
Thermal properties, 13
Thermal shock, 13, 76
Thermal stress, 76
Thermoforming, 48, 53
Thermoplastic polymers, 19, 29, 48
Thermosetting polymers, 19, 35, 38, 63
Three-directional reinforcement, 153, 154
Tobermorite, 166
Toughness, 8, 76
Tracheids, 190

Traditional ceramics, 71, 84, 103, 104
Transfer molding, 64, 65
Transition zone, 170
Translucent glass, 97
Transparency, 91
Tricalcium aluminate, C₃A, 165
Tricalcium silicate, C₃S, 165
Turbine rotors, 78

Ultimate tensile strength, 74, 75
Unidirectional, 133, 135
Unit cell, 80
Uranium oxide, 81, 82

Vacuum forming, 53
Veneer, 205, 206
Vessels, 190, 192
Vinyl linkage, 20, 21
Viscoelasticity, 6

Viscosity, 48, 98, 111
Vitreous, 91
Vulcanizing, 39, 40

Warpage, 198
Water, 168
Water/cement ratio, 168, 172
Wear, 12
Whiskers, 74, 127
Whiteware, 104
Wood composites, 205
Wood properties, 193
Wood pulp, 208
Wood structure, 190
Workability, 104, 168
Working range, 99

Zachariasen's rules, 93
Zirconia, 81, 82